TECHNIQUES FOR TECHNICAL COMMUNICATORS

Carol M. Barnum
Southern College of Technology

and

Saul Carliner

Macmillan Publishing Company
NEW YORK
Maxwell Macmillan Canada
TORONTO
Maxwell Macmillan International
NEW YORK OXFORD SINGAPORE SYDNEY

Editor: Barbara A. Heinssen
Production Supervisor: Publication Services, Inc.
Production Manager: Aliza Greenblatt
Cover Designer: Proof Positive/Farrowlyne Associates

This book was set in Palatino by Publication Services, Inc.,
and was printed and bound by Arcata Graphics.
The cover was printed by New England Book Components.

Printed in the United States of America

Macmillan Publishing Company
866 Third Avenue, New York, New York 10022

Macmillan Publishing Company is part of
the Maxwell Communication Group of Companies.

Maxwell Macmillan Canada, Inc.
1200 Eglinton Avenue East
Suite 200
Don Mills, Ontario M3C 3N1

Library of Congress Cataloging-in-Publication Data

Barnum, Carol M.
Techniques for technical communicators / Carol M. Barnum and Saul
Carliner.
p. cm.
Includes bibliographical references and index.
ISBN 0-02-306095-6
1. Communication of technical information. I. Carliner, Saul.
II. Title.
T10.B27 1993
601.4–dc20 92-10660
CIP

Printing: 1 2 3 4 5 6 7 Year: 3 4 5 6 7 8 9

Preface

Techniques for Technical Communicators is a collaborative effort between academe and industry. *Techniques* is also a collaboration among some of the best-known practitioners and researchers in the field. Finally, *Techniques* is a collaboration that capitalizes on the respective strengths of its editors. Carol Barnum has experience in publishing for the academic market and in teaching upper-division and graduate courses in technical communication. Saul Carliner has experience in developing many types of technical documents and knows many of the contributors through his participation as a conference organizer and journal editor in professional societies. Together, we had a good sense of what the market needed. We targeted a gap in the market and designed *Techniques for Technical Communicators* to fill it.

Intended for advanced students of technical, scientific, and professional communication programs, as well as working professionals and those interested in entering the field, *Techniques for Technical Communicators* takes you through the process of preparing, writing, editing, and evaluating documents, whether for print or other media. Research based, but practical in its application and approach, *Techniques* presents the skills needed to succeed as a professional in the field of technical communication. But *Techniques* also goes a step beyond to explore such key issues as working with people, producing information in multimedia, testing document usability, and understanding the legal and ethical aspects of protecting your work and using the work of others. An understanding of these issues will help you grasp the full range of activities of working professionals in our field.

We are fortunate in having among our contributors many of the leaders and rising stars in technical communication. The contributors speak to you in their own voices; and although their styles may vary, their delivery is always lively, practical, and well researched.

All the chapters have been commissioned specifically for this book. Each begins with an overview of the topic under discussion and concludes with a summary highlighting the main points covered. To reinforce the practicality of the information presented, contributors use illustrations, case studies, and extended examples. Full bibliographies append each chapter, along with lists of terms introduced. For easy reference, these terms are italicized when introduced in the text.

We've organized the chapters to focus on the process of technical communication, breaking the process into four distinct phases.

- Techniques for planning information
- Techniques for designing information
- Techniques for editing information
- Techniques for verifying and protecting information

Although, as we point out, these phases are not necessarily linear, we present the process as it should and often does take place in industry and business.

Students and professors of technical, scientific, and business communication, as well as working professionals, should find ideas that stimulate, material that's pertinent to a wide variety of courses in upper-division and graduate programs, and strategies that work.

As our book is diverse, so is the field it describes. Not surprisingly, *Techniques* suits a wide variety of courses with such titles as Document (or Information) Design, Issues in Science and Technology, Writing for the Sciences (or Professions), Advanced Technical Writing, Technical Editing, and Instructional Systems Design, to name a few. Because the professors who teach these courses are generally well versed in their subject area, *Techniques* does not suggest writing assignments or exercises, thus leaving professors free to use the text as a basis for designing a course of their choosing.

We are indebted to many people for their help in bringing this book to you. First of all, our thanks go to Barbara Heinssen, Senior Editor at Macmillan, who recognized the possibilities even in the early stages and who carefully shepherded the project through all its phases. Thanks also go to Rachel Wolf, Assistant Editor, Macmillan; Aliza Greenblatt, Freelance Manager, Macmillan; and Melissa Madsen Blankenship, Senior Production Coordinator, Publication Services, for all their able assistance. We would also like to acknowledge the help of our reviewers: Virginia A. Book, University of Nebraska—Lincoln; Roger Friedman, Kansas State University; Dean G. Hall, Kansas State University; Mark Jarvis, Cowley County Community College; M. Jimmie Killingsworth, Texas A & M University; Carolyn R. Miller, North Carolina State University; James Porter, Purdue University; and Donald E. Zimmerman, Colorado State University. Their excellent comments and suggestions to us have made this a better book. Finally, our thanks go to our contributors who faithfully worked to meet our deadlines, willingly revised as needed, and who have always believed in and supported this project. We are grateful to all of them.

C. B.
S. C.

Foreword

M. Jimmie Killingsworth
Director of Writing Programs
Texas A&M University

In the 15 years that I've worked in technical communication, the field has grown tremendously, outstripping the old textbooks and how-to articles that were geared mainly to engineering writers in the aerospace and defense industry. The global explosion of information, the advent of the personal computer and desktop publishing, the expansion of electronic media for communication, the growth of the service sector, the increasing sophistication of research in composition, and the evolution of extensive programs in colleges and universities from what used to be, at most, a single course in technical writing—these changes have sent teachers and students looking for up-to-date readings that offer practical instruction without watering down the knowledge that researchers and technical communicators have gained over the last couple of busy decades.

Techniques for Technical Communicators comes closer than any other book I know to meeting these needs. Carol Barnum and Saul Carliner have gathered chapters by authors who are, like the editors themselves, experienced teachers, researchers, and practitioners sensitive to the needs of both novices and more advanced students in technical communication. The greatest compliment I can pay them, perhaps, is that they take their own definition quite seriously. If technical communication is "the process of translating what an expert knows for an audience with a need to know," then *Techniques for Technical Communicators* is not only a book about technical communication, but also an excellent example of it. In this book, some of the most respected experts in the field address an audience of users who must put new knowledge to work without confusion or undue delay. These experts are certainly up to the challenge. They give us cutting-edge information and highly technical instructions in a form both easy to use and personally engaging. The authors practice what they preach, and what they preach is well worth hearing.

Both beginners and veterans in technical communication will appreciate the authors' efforts to make information accessible. The fact is, technical communication has grown rather technical in recent times. Building upon cognitive science and observational research in industry, researchers have made special progress in two areas: the study of the communication process and the study of document design. These two topics form the theoretical backbone of the practical advice given in this book.

The chapter organization as a whole reflects the key stages of the process by which technical communicators plan, develop, revise, edit, produce, and review documents. Drawing upon practical experience as well as recent theory and research, the authors reveal the overall movements and the more subtle contours of the document-development process. They allow us to enter the process through a variety of roles—those of the writer, the artist, the editor and the manager. And they give us the knowledge and confidence to undertake all of these roles in our own efforts to produce effective and usable documents.

While the process approach favors the perspective of the document producer, the concept of design focuses on the receiving end of the transaction. The principles and techniques of document design keep the writer's mind and the artist's eye turned always in the direction of the reader. The typical reader in technical communication, an active information *user,* requires that documents, like all engineered products, make application easy and help to solve problems in the working world. From Chapter 1 (by Janice Redish, a true pioneer in document design) through William Horton's fine chapter on visual communication to the penultimate chapter by Ann Hill Duin on usability testing, *Techniques for Technical Communicators* keeps the focus on the reader by explaining and applying the best principles of verbal rhetoric and graphic design.

Finally, I must mention the book's remarkable economy. Many of the authors give short courses that quite thoroughly cover topics that have consumed whole books—indeed whole shelves of documents. Saul Carliner's overview of presenting verbal information and Fern Rook's summative chapter on grammar and style come to mind as obvious examples. These well-crafted and thoughtful chapters attain brevity without sacrificing depth and breadth of coverage—an accomplishment to which we should all aspire.

All told, few books provide in so short a space such a wealth of timely, dependable, research-based, experience-tested information. I am pleased to introduce and recommend this major achievement to students and teachers in technical communication.

Brief Contents

Detailed Contents

INTRODUCTION

Our society is in the midst of an information explosion. According to Mike Braun (1991), Vice President for Multimedia at IBM, "The world's information base doubles every three years" (p. s.s.1). In practical terms, an information explosion means that people have more news sources, more magazines, more television channels, more work-related information, more consumer information—in short, more information than they need or can manage.

As people have become increasingly burdened by the ever-growing supply of information, a profession of information specialists, called *technical communicators*, has emerged to manage this growing information supply. Technical communicators lighten the information load on others by preparing information for effective, efficient use.

Not surprisingly, the growth in the technical communication profession has mirrored the growth of information in our society. In 1984, membership in the Society for Technical Communication (STC) numbered 5,415. Today, that number tops 16,000, with the actual number of practitioners estimated at more than 100,000.[1] As recently as 15 or 20 years ago, few technical communicators were academically trained in their discipline because few programs were available. Now, new programs are added every year at all levels of academe. Although the Council for Programs in Technical and Scientific Communication and the STC have attempted to track and document the number of programs, no comprehensive list exists. The Society for Technical Communication estimates that 160 programs are currently in place.

1. These and other data about the numbers of technical communicators and the kinds of work they perform derive from The Society for Technical Communication's Membership Profiles.

Although the number of technical communicators increases, there remains some confusion about the field itself. This book explores both the nature of the field and the techniques of its practitioners.

WHAT IS TECHNICAL COMMUNICATION?

Technical communication is a broad and diverse subject. One way of defining it is to look at the activities of technical communicators. For 60% of technical communicators, technical communication means the practice of writing software manuals and online information. The other 40% work in many other industries and perform many other tasks.

Fourteen percent of technical communicators design, write, and edit training materials. Training covers a variety of subjects, including "hard" subjects like manufacturing, machines, and metrics, and "soft" subjects like managing projects and mastering negotiation skills. These training materials include workbooks, videos, computer-based tools, and multimedia (a combination of video, audio, and computer graphics). The designing aspects of the task do not refer solely to the visuals to support text, but more broadly to the approach used in sequencing information, determining the best way to present it, and choosing the best medium for the message.

Technical communicators also research, write, edit, and produce marketing and advertising materials. These materials include brochures, catalogs, and advertisements in all media, as well as sample diskettes distributed at trade shows and multimedia kiosks located in airports and shopping malls. Annual reports, speech writing, and multimedia presentations at conventions are also examples of this type of technical communication.

Other technical communicators report recent scientific and technical discoveries. This information might be communicated through articles in research journals and the trade press, through presentations at professional meetings, and through exhibits at trade shows. Still other technical communicators work in the insurance industry, in banking and finance, in health services, in government, and in the military and defense industries. In other words, technical communicators work in every industry where information needs to be exchanged from one group to another.

A DEFINITION OF TECHNICAL COMMUNICATION

Despite the diversity within the field of technical communication, technical communicators can be expected to perform certain common tasks, which constitute a definition of the field:

> Technical communication is the process of translating what an expert knows for an audience with a need to know.

Despite its apparent simplicity, this definition encompasses the key concepts underlying effective technical communication, regardless of the subject matter, format of the message, or medium used. We focus our discussion on three key concepts in the definition: *audience, translation,* and *process,* which we discuss in the remainder of this chapter.

Technical Communication Is Intended for a Specific Audience

Technical communication is characterized first by the need to identify and understand the intended audience for the information. Some audiences are unified; that is, they share common characteristics. Consider first-time users of a scanner, a piece of equipment that is connected to a computer to "read" printed materials into a particular computer system. These first-time users have no previous experience with the scanner, but they do share basic familiarity with the computer that will store the scanned documents.

Other audiences are diverse; that is, they share few common characteristics. Consider the readers of a feasibility study. They include managers in different divisions of an organization. In this case, while all of the audience is managers, the expertise of each is varied, not only with regard to the subject of the report, but also with regard to specific sections.

In both examples, however, the audience has a *need to know.* Only those with a need to know will be using the information, and the purpose for which they'll be using the information determines *what* they need to know.

Thus, technical communication is audience-based, not writer-based: it focuses on the needs of the audience rather than on the wants or desires of the writer. Not surprisingly, the roots of technical communication are in the field of rhetoric, particularly persuasion, rather than literature or creative writing. Decisions about purpose, style, format, information, approach, and medium are all based on good audience analysis.

Technical Communication Is Translation

Technical communicators must take complicated subject matter, easily understood by subject-matter experts, and "translate" it into a language, a format, a style, and a tone that can be easily understood by nonspecialists.[2] That is no easy task, for translation requires gathering information from one or more groups and passing it to another group in a different form. It requires recognizing jargon—the specialized vocabulary of one group—and reducing it to terms and expressions that can be understood by those outside the group. It means asking the right questions to get enough information from the subject-matter experts to make a translation possible. It means playing the part of the users to anticipate the problems they might encounter and questions they might ask. And it means learning enough about the subject to write accurately and completely about it. Thus, technical communicators must combine their knowledge of the audience with their ability to translate as they engage in the process of technical communication.

Technical Communication Is a Process

The process of developing technical documents has many steps: brainstorming, outlining, drafting, editing, testing, and revising. These steps are rarely as linear as they may appear on a project manager's master schedule. But the steps can be described as a process.

This process, from inception to completion, can be represented as follows:

1. Understanding the readers
2. Organizing the information
3. Planning the project
4. Forming groups to work with other people
5. Drafting the information verbally and visually
6. Producing information by
   - Designing it for print or online
   - Formatting it for multimedia
   - Considering matters of grammar and style
   - Editing for consistency
7. Evaluating the usability of documents
8. Protecting information through copyrights and professional ethics

---

2. In using the term "translation" in this specialized way, we in no way mean to detract from the literal act of translation from one language to another, which is an important type of technical communication.

Although technical communication projects follow a steady progression, such as the one presented above, some steps may be compressed when deadlines must be met and other steps may be repeated when the discoveries made in testing require changes in verbal or visual presentation. Nonetheless, the recognition of this process is the natural starting point for planning, creating, and composing any form of technical communication, the result of which is information in a usable form.

Thus, our definition of technical communication unifies a seemingly diverse field: technical communicators need to understand

- Who the intended audience is
- How to function as translators in preparing information
- How to manage the production of information

TECHNIQUES FOR TECHNICAL COMMUNICATORS

With so many steps in the process, so many skills to master, and so many technologies to use, you may wonder where to begin learning about the field. The ever-widening range of technologies available to technical communicators—hypertext, multimedia authoring systems, CD-ROM, to name but a few—suggests that the most important skills are mastering these tools.

But they're not. Tools change with time, but the basic skills of communication—planning information, designing, writing, editing, testing, and working with other people—remain the same. For that reason, this book focuses on these skills. It is specifically intended for advanced students of technical, scientific, and professional communication programs, and for practicing professionals.

Techniques for Technical Communicators explores the process of developing technical information. Each section of the book presents practical, research-based techniques for performing the steps in this process:

- "Techniques for Planning Information" explains how to understand your readers and organize information for print and online media. This section also prepares you to work on a technical communication project by planning the budget and schedule and learning to work collaboratively with other members of the project team.
- "Techniques for Designing Information" explains how to present many types of information through words and pictures.

- "Techniques for Producing Information" explains formatting and design considerations for print, online documents, and multimedia, plus considerations of grammar, style, and editorial consistency.
- "Techniques for Verifying and Protecting Information" explains how to evaluate the usability of documents and protect your work through copyrights and through professional ethics.

Although technical communicators need to be adept at all the techniques of technical communication, most tend to specialize in one or two areas associated with the steps in the process. Few, if any, are expert in all these areas, Nor are we. That is why we have invited experts in each step of the process to write chapters in their areas of specialization. Like us, our contributors represent expertise from both industry and academe: the practical and the theoretical. We invited our contributors to speak to you not only in their areas of expertise, but also with their own voices. We hope that you will find the variety stimulating and challenging. We also hope that this book will illustrate the diversity of the field of technical communication and that it will stimulate your thinking about your place in it.

ABOUT THE CONTRIBUTORS AND THEIR CHAPTERS

Chapter	About the Chapter	About its Author
Understanding Readers	Explores the basic characteristics of readers that have been identified through research and suggests strategies you can use to most effectively reach your readers.	*Janice C. Redish* is a Senior Fellow with the American Institutes for Research and the former Director of its Document Design Center in Washington, DC. In her position, she has the opportunity to perform research and apply it to actual publications, and she has received the 1988 Best Article Award from *The Technical Writing Teacher,* the Golden Plaque Award from the National Computer Graphics Association, and several Society for Technical Communication publications awards for her work. Redish is a widely published author on the readability and usability of technical information. She holds a Ph.D. in Linguistics from Harvard University.

(*continued*)

The Design Draft—Organizing Information

Explains how to "design" a document. Specifically, this chapter explains the importance of organization to a document, how to set the goals for a document, considerations for selecting a communication medium, how to translate the goals for a document into a design (plan for presenting information), and how to present the design to others.

Heather Fawcett is a principal with Information Design Solutions in Guelph, Ontario. She is a co-developer of numerous workshops on technical documentation, including Laying the Foundations, which explores design draft issues. Fawcett also worked on the Oxford English Dictionary project at the University of Waterloo, which developed tools to access the electronic version of this dictionary. She has research and practical experience in user interface design and hypertext. Fawcett holds a B.A. in English and Applied Studies from the University of Waterloo.

Samuel Ferdinand is a principal with Information Design Solutions in Brampton, Ontario. He is a co-developer of numerous workshops on technical documentation, including Laying the Foundations. Formerly a manager with Computer Logics in Rexdale, Ferdinand has experience in project management, documentation design, and document evaluation. He holds a B.Tech in Mechanical Engineering from Ryerson Polytechnical Institute and a Certificate in Technical Writing from Durham College.

Ann Rockley is a principal with Information Design Solutions in Stouffville, Ontario, and a co-developer of numerous workshops, including Laying the Foundations. Rockley, a former manager of instructional design with Apple Canada, has extensive experience with hypertext and with designing, developing, and managing documentation and training projects. Rockley is a past president of the Toronto chapter of the Society for Technical Communication (STC) and was general manager of Technicom 89. She holds a B.A. in Creative Writing, with a minor in Science, from York University.

(*continued*)

Chapter	About the Chapter	About its Author
Planning and Tracking a Project	Presents a six-step process for determining the scope of a project, estimating its cost, scheduling work, and tracking the project once it begins.	*James Prekeges* is an instructional designer with Microsoft Corporation in Bellevue (Seattle), Washington. He is the author of several papers on project planning and is a frequent speaker at Rensselaer Polytechnic Institute's Technical Writer's Institute. Prekeges is a past president of the Puget Sound chapter of the STC. He holds a B.S. in Technical Communication from the University of Washington.
Working with People	Examines the collaborative nature of technical communicators' jobs. This chapter explains what collaborative writing is, describes the various aspects of collaborative writing, describes the role of personality and conflict in collaborative writing and small group communication, and presents techniques for successfully working with other people.	*Carol M. Barnum* is a Professor of English at Southern Tech in Marietta (Atlanta), Georgia, where she teaches graduate and undergraduate courses in technical communication, including courses in small group communication. She has served as Acting Director of Southern Tech's master's program in technical communication, the McFadden Visiting Professor of Technical Writing at Lehigh University, and as an exchange professor in China. Barnum serves on the international Board of Directors of the STC and as editor of the "Practicalities" column in *Technical Communication*. She holds a Ph.D. in English from Georgia State University.
A Way with Words—Presenting Information Verbally	Describes tips and techniques for presenting various types of information, including techniques for instructing, informing, and persuading readers.	*Saul Carliner* is a Marketing Programs Administrator with IBM in Atlanta, Georgia, where he coordinates marketing communications programs for the Customer Education Business Unit. He has published several award-winning publications and is a frequent speaker on the future of technical communication. He is Associate Editor for Information Design of *Technical Communication*, has served on the committees of four STC Annual Conferences and as past president of the Atlanta Chapter of the National Society for Performance and Instruction. He holds a

(*continued*)

		B.A. in Economics, Professional Writing, and Public Policy, with a minor in Administration, from Carnegie Mellon University and an M.Ag. in Technical Communication from the University of Minnesota.
Pictures Please—Presenting Information Visually	Describes tips and techniques for using visual tools, such as charts, lists, and illustrations to communicate information.	*William Horton* is president of William Horton Consulting, a Huntsville-based consulting firm. Author of *Illustrating Computer Documentation* and *Designing and Writing Online Documentation,* Horton is an internationally known expert on the visual communication of technical information. Horton is a Fellow of the Society for Technical Communication, the editor of "The Wired Word" in *Technical Communication* and a registered engineer. He holds a B.S. in Mechanical Engineering from MIT and an M.S. in Computer Science from the University of Alabama at Huntsville.
Design That Delivers—Formatting Information for Print and Online Documents	Explains the role of page and screen design in successful documents and presents design tools you can use throughout documents and on individual pages and screens.	*Martha Andrews Nord* is a Professor of Management and Technical Communication at Vanderbilt University in Nashville, Tennessee, with a joint appointment in the Graduate Business School and the School of Engineering. She is also a principal in Nord Consultants and is National President of the Association of Professional Writing Consultants. She has performed research on document design, graphics, and teamwork in technical communication, and is a frequent speaker on these subjects. Nord holds a Ph.D. in English from Vanderbilt University.
		Beth Tanner is President of Tanner Corporate Services, Inc., in Nashville, Tennessee. Tanner has a strong background in document design and project management and currently consults with *Fortune* 500 firms regarding technical communication. She holds an M.B.A. from Vanderbilt University and a B.A. from Brown University.

(*continued*)

Chapter	About the Chapter	About its Author
Presenting Information through Multimedia	Explains what multimedia, hypermedia, and hypertext are, when you should use these media, and how to design and develop documents to be presented in these media.	*Francis D. Atkinson* is a Professor Georgia State University in Atlar and director of its graduate progra in Instructional Technology, whe he teaches courses in hypermedi message design and instructiona design. He is the coauthor of *Instructional Media for Training and Education* and was Executive Director of Learning Initiatives. I holds a Ph.D. in Instructional Te nology from Syracuse University.
Remembering the Details—Matters of Grammar and Style	Explains the differences between grammar and style and suggests a framework for making stylistic choices in your own writing.	*Fern Rook* is a retired Professor o English from Arizona State University and the former autho of the popular column "Slaying the English Jargon" in *Technical Communication,* in which she described grammatical issues from the perspective of technical communicators. Rook is a Fellow of the Society for Technical Communication. She holds an M.A. in English from Arizona State University and a B.A. in English from the University of Colorado.
Sweat the Small Stuff—Editing for Consistency	Explains the importance of consistency for clear technical communication and methods for identifying and controlling consistency through editing.	*Vee Nelson* is President of V. Nelsc Associates in Decatur (Atlanta), Georgia. An experienced editor and former professor at Morehouse College, Nelson now specializes in the development of instructional materials. Nelson is a former president of the Atlanta chapter of the National Society fo Performance and Instruction. Nelson holds a Ph.D. in English from Georgia State University.
Test Drive—Evaluating the Usability of Documents	Describes the role of usability testing in the development process and explains how to plan and run usability tests.	*Ann Hill Duin* is an Associate Professor of Rhetoric at the University of Minnesota in St. Pau She is a widely published author on usability and document desigr Duin received the 1989 EDUCOM Award for the best curriculum innovation in teaching writing. She received the award for a computer-based curriculum that taught collaborative writing

(continue

		through telecommunications. An award-winning educator, Duin has also developed innovative uses of computers to teach writing. Duin holds a Ph.D. in English Education from the University of Minnesota.
Protecting Your Work—Professional Ethics and the Copyright Law	Explains the purpose of the copyright laws, how to use copyrighted material in your work, how to copyright your material, and the ethical considerations associated with the copyright laws.	*Jacquelyn L. Monday* is the Chief of the Publications Section of the Louisiana Geological Survey, where she oversees the production of the Survey's formal publications and manuscripts submitted to scientific journals. Monday previously was an editor for the University of Colorado's Institute of Behavioral Science and a policy and regulations specialist for the U.S. government. Monday holds a B.A. and an M.A., both in Geography, from the University of Colorado. *Mary C. Hester* recently began studies in law following a career as a freelance editor and writer and a part-time instructor at Louisiana State University, where she taught technical and business writing. Hester holds an M.A. in English from Louisiana State University.

REFERENCES

Braun, M. (1991). "Personalizing the information revolution." *T.H.E. Journal, 19* (2), s.s.1 (special supplement).

PART I

TECHNIQUES FOR PLANNING INFORMATION

CHAPTER 1

UNDERSTANDING READERS

Janice C. Redish

American Institutes for Research, Document Design Center

Understanding your readers is critical to producing a useful document, and teachers of technical communication regularly emphasize the need for audience analysis. To understand their audiences, technical communicators often ask specific questions about their readers' jobs, education, age, and needs—important information for the specific document. Equally important is understanding more generally how people act as readers, that is, how they work with documents. That more general understanding is the focus of this chapter.

The chapter has four major sections, each devoted to a critical aspect of how readers work with documents:

1. Readers decide how much attention to pay to a document.
2. Readers use documents as tools.
3. Readers actively interpret as they read.
4. Readers interpret documents in light of their own knowledge and expectations.

In each section, I discuss both relevant research and practical implications for technical communicators.

READERS DECIDE HOW MUCH ATTENTION TO PAY TO A DOCUMENT

Just writing a document isn't enough to ensure that people will read it. Reading is a voluntary act; people don't have to do it.

What the Research Says

Wright, Creighton, and Threlfall (1982) found that people have definite opinions about when instructions should be necessary. They asked the subjects in their study, who were typical adult consumers, to imagine that they had just bought products ranging from cake mix to bleach to a digital watch to a washing machine. There were 60 products on the list, all with made-up brand names. Wright and her colleagues asked these consumers how much of the instructions they would read for each product.

The results showed both that consumers have strong feelings about the types of products for which they would and would not read instructions and that many consumers agree on the products for which they did and did not expect to need instructions. For example, they wouldn't read instructions for a TV because they expect to only have to plug it in and switch it on. You can't assume that, just because you wrote a document, people will read it.

People's beliefs about the need for instructions are difficult to overcome. Wright (1981, 1989) relates this example: When a British candy company put out a very strong peppermint that was not meant to be eaten like candy, it put a caution notice on the package. The caution notice didn't work. The product looked like candy; it was sold with candy; and people bought it and ate it like candy—and got sick. The company was surprised, but it should not have been. People don't look for instructions on a package that seems to contain candy. As a technical communicator, you need to be aware of readers' expectations and of potential conflicts between the message that the packaging sends and the messages in your illustrations and written words.

People may decide not to deal with documents even when it means giving up something they deserve and need. For example, in one of our early studies, my group at the American Institutes for Research found that poorly written forms and notices are overwhelming to many readers (Rose, 1981). People miss deadlines, don't turn in applications, and don't pursue incorrect refunds or benefits statements because the documents are just too difficult to understand and use. You cannot assume that people will work their way through a document just because it is important. If the organization, language, and design make the document too difficult, many readers will give up.

In a 1990 study, we found similar results on a much larger scale. We invited readers of *Modern Maturity,* the magazine of the American Association of Retired Persons, to answer questions about their experiences with forms and notices. We asked them to select one form or notice that they had recently dealt with and to answer questions about that document.

Of the more than 3,800 people who responded, 34% said that they had lost money or benefits because the document was too difficult to understand or fill out. Of the 474 respondents who wrote about forms that they were trying to use to apply for credit or to buy something, 29% gave up before finishing the process, and 30% stopped using the service or organization (Bagin & Rose, 1991a, 1991b).

These readers lost out because they decided not to deal further with the document, but the businesses that were trying to communicate with these people also lost. They lost potential customers and future sales, perhaps because they did not realize that, even when people start out wanting whatever the document covers, they may give up on both the document and the company if the document is too difficult.

Writing so that people read a document is not restricted to those who communicate with the general public. In the workplace, your technical documents compete with many others for readers' attention. The most salient characteristic of readers in the workplace is that they are busy and don't have time to waste on dense documents. Consider your own experiences as a reader. How many documents have you glanced at and put aside to read later because they looked too dense to deal with immediately? How many of those have you never really read?

Even sophisticated, technical readers of technical documents want to be able to open a document, find what they need quickly, understand it easily, and have the most important information stand out visually on the page or computer screen.

Readers are continually deciding, consciously or subconsciously, whether more time and effort with a document is worth the additional benefit in learning or understanding. I've found it useful to think of this as *satisficing*, to borrow a term from the work of the well-known economist Herbert Simon (1976). Simon used *satisficing* to describe what administrators do when they "look for a course of action that is satisfactory or 'good enough.'" They act without exploring all the options. They use "rules of thumb that do not make impossible demands upon [their] capacity for thought" (Simon, 1976, p. xxx).

Readers working with technical documents and products also satisfice. They skim; they skip; they read just enough to reach a personal level of satisfaction with their new knowledge or until they reach a personal level of frustration with the document or product. They may return to the document later for more information or to try again, which is another reason why helping readers to find what they need quickly is so critical in technical documents.

Many people continue to do tasks in less than optimal ways because the effort to figure out a new way seems too great. How many functions of your own VCR, microwave oven, telephone, or word processor do you use? Do you ever want to do more or do something differently—for example, record one program on your VCR while you're watching another program, address an envelope using the computer, or place a conference call—and yet not do it because it's too much trouble to read the documentation? If so, you're satisficing.

In many cases, satisficing is the only way that readers can cope with their work environment. For example, subject matter specialists who read proposals for grant agencies may have 10 to 12 proposals to read in five hours. That's 20 to 30 minutes per proposal. Much as these readers want to be fair and give each proposal all the points it deserves, they have to satisfice. They don't have the time to hunt out information that may be hidden under an obscure heading. They don't have the time to untangle a wordy and convoluted plan. Proposal writers who have made it easy for these readers to find and understand the relevant information are going to get higher scores.

What You Can Do

What implications do these facts about readers have for technical communicators? One is that you should consider whether you are writing the best type of document to help your readers. For example, given the fact that many people were overwhelmed with a particular form, Carolyn Boccella Bagin of the American Institutes for Research turned the form into a series of letters (Bagin, 1988). As I'll discuss later in this chapter, people are more comfortable with letters than with forms. (Of course, not every form would be better as a letter, yet most forms you see could be vastly improved by a technical communicator who understands readers and knows techniques of good document design.)

Another implication is to deal head-on with the reality that busy readers may not be willing to read much. For example, given the problem of how to get executives to start using electronic mail, Joanne Landesman of the American Institutes for Research created a 12-panel card of *Mini-Lessons*. The *Mini-Lessons* card started with a bulleted list of what executives could do through the electronic mail program. It then gave the information these users needed to get into the program and showed them—with step-by-step instructions and a picture of the screen—how to do the most basic tasks. Each task was on a separate panel of the fold-out card. That's all these executives were willing to work with, and it was enough to get them going.

A third implication is to consider how your readers will use the document. Will they primarily browse through it, stopping to read what catches their interest? Will they search it for one specific piece of information? Will they read it from beginning to end? Will they return to it later to look for something they had read earlier? The answers to each of these questions imply specific techniques for organizing and designing documents, techniques that are discussed in the later chapters of this book.

READERS USE DOCUMENTS AS TOOLS

In a work of fiction, the value of the work may be intrinsic. In a workplace document, however, the value is almost always extrinsic, that is, external to the document itself. As a technical communicator, you want readers not just to read the document, but to use it. Consider these situations:

When you write this type of document:	**You may want readers to do this:**
Engineering specification	Build the bridge properly
Feasibility study	Make a wise decision
Instruction manual	Use the product effectively and efficiently
Memo about a meeting	Come to the meeting on time and prepared

Progress report	Continue to fund your project; agree to changes
Proposal	Fund the project
Report of an experiment	Use the same method; build on your results

What the Research Says

In the workplace, people most often "read to do" rather than "read to learn," although the opposite is true in school (Sticht, 1985; Mikulecky, 1981). In school, the main point of reading is to learn information that you will have to remember at some future time. In the workplace, many documents, from memos to reports to manuals, are used at the moment they are read to serve an immediate need. Busy readers don't generally study documents. Instead, they scan them to find the critical information and to act on that information. If they need more information, or even the same information later, they can look it up again. The working person's primary job is not usually to read documents. It is to do something, such as get a bridge built, put people on the payroll, or decide which machine to buy. Even technical communicators don't get paid for the amount they read; they get paid for writing successful communications.

Because many technical documents are tools that people use to do their jobs, they want to get in, get the information, and get out of the document. Wright shows this clearly in her model of how people use technical documents (Figure 1.1).

Wright's View of Using Technical Information

Before reading

- Formulate question—but not necessarily a precise question
- Find the location of a potential answer

During reading

- Comprehend text—but reading may be highly selective
- Construct action plans or use information to make a decision

After reading

- Execute action plans or implement decisions
- Evaluate outcome of actions or decisions

FIGURE 1.1 How People Use Technical Documents (Wright, 1987, p. 340)

Wright (1988) says that the questions people ask of technical information are problem-driven ("Why didn't that work?") or task-driven ("How do I do that?"). They are not usually system-driven ("How does it work?"). Wright's view of how people use technical documents is very similar to Donald Norman's view of how people use such objects as lamps, doors, coffeepots, VCRs, and computers (Norman, 1988). People use objects and the technical information about them in actual situations. The objects and information serve as the means for accomplishing plans and goals that are relevant to those situations. People focus on the plans and goals, not on the objects or technical documents themselves.

Research in the 1980s showed that Wright's model of how people use technical documents is clearly true for manuals. Many people use a manual only as a last resort. They don't read it; they go to it when they have a problem to solve. In a study of users trying to locate library books through a computerized catalog, Sullivan and Flower (1986, pp. 170–171) discovered the following about how people used the manual for the catalog:

- No one carefully read more than two sentences of the manual at a time.
- Most people began to use the product before they turned to the manual.
- The people in the study used the manual only when they were not successful in achieving their goal.
- Most did not read the introduction first; nor did they read all of it, even though it was only three short paragraphs.
- Most did not read any section in its entirety.

Even documents that require a careful reading initially, which may be the case with many scientific papers and technical reports, are often used again later for reference. At that point they become just like the manual in Sullivan and Flower's study. Readers go to them looking for a specific piece of information or the answer to a specific question—and these readers follow Wright's model.

Not everyone, of course, will act like the people in Sullivan and Flower's study. One of my points in the next section of this chapter is that each reader approaches documents differently. In usability laboratory studies at the American Institutes for Research, we have found that some people are readers and others are nonreaders. Some people look up information, use online help regularly, and are happy with explicit instructions, for example, in tutorials. Others are risk takers who prefer to work with a product on their own, who seldom look in a manual or in online help, who would rather explore than work through a tutorial (Redish, 1988).

These findings echo a study by *PC Magazine* (1988) in which more than 1,200 people responded to questions about learning to use new products. Of the six choices for "How do you learn new software?" only two got significant numbers of responses: "read the manual" and "experiment on your own," each of which came in at 45%. In the same study, *PC Magazine* asked, "What's your greatest obstacle in learning software?" Fifty-nine % chose the option "poorly written manuals." No other response received more than 15% of the votes. Many people, however, also commented that "lack of time" was a major problem.

What You Can Do

What are the implications of the fact that busy people use documents as tools? One implication is that because readers want help in finding the information they need quickly, good access tools are critical. The table of contents needs to match the topics that readers will be looking for. The index needs to include words that readers bring to it. (Also see the discussion in the fourth section of this chapter on the role of a good table of contents in giving readers an overview of the document.) The later chapters of this book, as well as Redish, Battison, and Gold (1985), offer many suggestions for organizing information so that it is easily accessible to readers.

Another implication is that readers want to grab information off the page quickly—often without sustained reading. For busy readers, you need to design pages that can be skimmed, scanned, and searched easily. To put white space onto a page and to chunk (separate) the information into usable pieces, you can use headings that stand out from the text; lists that set off steps or items with numbers or bullets; relevant visuals, such as tables, charts, and illustrations; and short paragraphs.

A third implication is that you have to understand how people are going to work with the document. Will people use it primarily to "read to learn"? If so, structures that promote recall, such as summaries, help readers. Will people use it primarily to "read to do"? If so, structures that promote action, such as numbered lists of steps, help readers.

Elsewhere I have suggested that some documents in the workplace, primarily computer tutorials and users' guides, must serve a hybrid purpose of "reading to learn to do" (Redish, 1988). Readers come to these documents wanting to use them like "reading to do" materials; they are eager to get on with their work. However, they will get on better with their work in the future if they also learn from the materials. Carroll and Rosson (1987) call this the "production paradox." Chapters 2 and 5 explore ways to develop documents that serve many uses.

READERS ACTIVELY INTERPRET AS THEY READ

Meaning does not reside in the text of a document; it exists only in the minds of communicators who produce documents and readers who use documents. Because each reader is an individual with his or her own knowledge, interests, and skills, a text can have as many meanings as it has readers. As Gopen and Swan (1990) write, "We cannot succeed in making even a single sentence mean one and only one thing; we can only increase the odds that a large majority of readers will tend to interpret our discourse according to our intentions" (p. 553).

What the Research Says

Early models of reading and writing that fixed the meaning in the text do not account for the dynamic nature of the reading process. Figure 1.2 shows the traditional, static model of reading and writing.

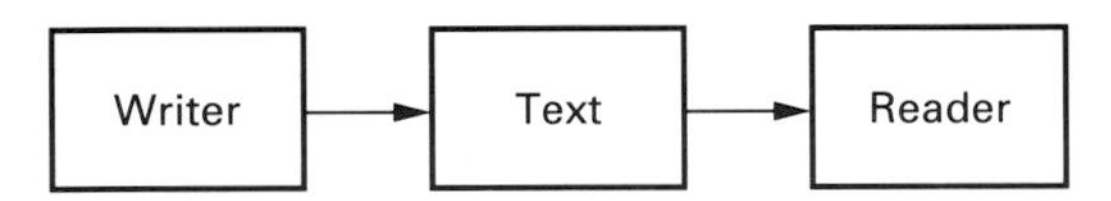

FIGURE 1.2 An Overly Simple, Static Model of Reading and Writing

A more appropriate model would show the complexities of the interaction, including the characteristics of individual readers and the characteristics of the document, all of which influence the reader's interpretation of the text. Figure 1.3 is an example of such a model from a 1982 paper by Holland and Redish.

As this figure shows, each reader brings his or her own needs, motivations, expectations, knowledge, and style to the interaction. Readers use documents to carry out tasks. To accomplish these tasks, readers have to form an internal representation of the document and select strategies for making sense of the document. That is, they have to figure out when, how, and why to use the document; whether and when to browse, search, read, or study; and how to carry out these different types of interactions with the document.

The Holland and Redish model in Figure 1.3 shows that these representations and strategies are influenced by the document's purpose and its place in an institutional or social setting as well as by the internal characteristics of the document. As the readers formulate representations and strategies, they are also influenced both by the content of the message in the document and by the ease or difficulty of using the document, that is, by its organization, format, syntax, and vocabulary.

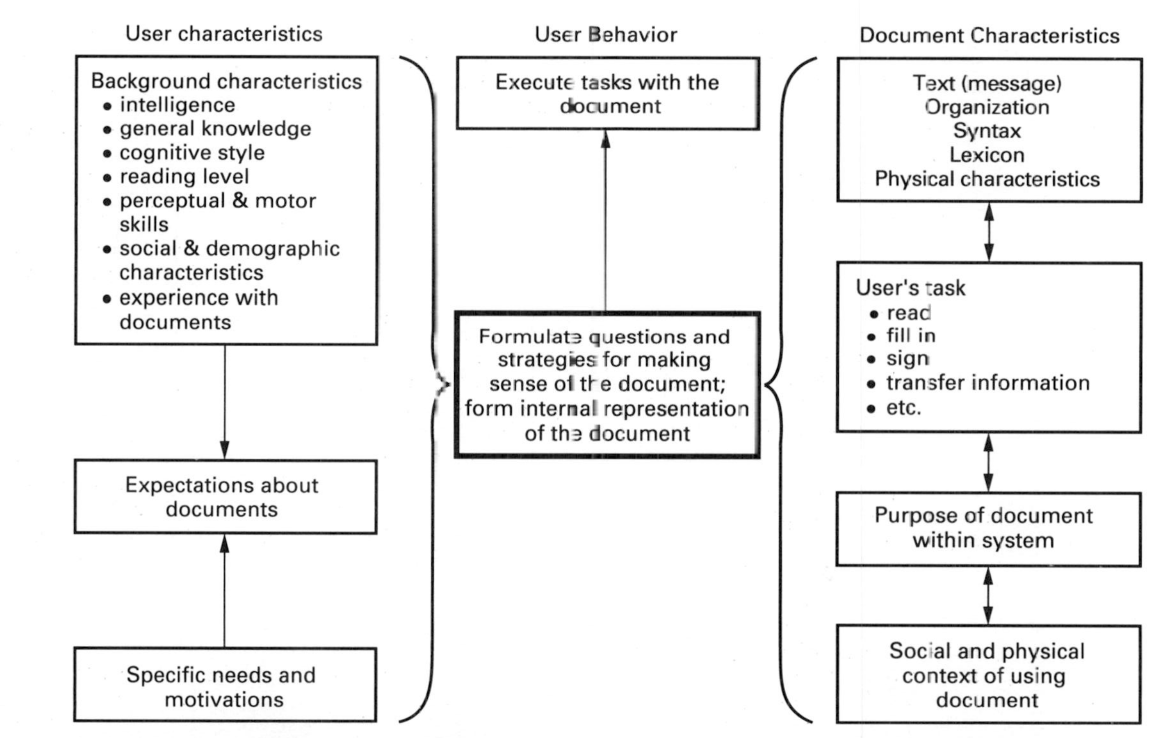

FIGURE 1.3 **A Model of the Interaction Between Readers and Documents (Holland & Redish, 1982, p. 206)**

A more recent model by Linda Flower, shown in Figure 1.4, adds yet another dimension to Holland and Redish's picture. Flower points out that the interpretation between person and text happens for both the communicator and the reader (Flower et al., 1990). Each is influenced by his or her own social context, goals, knowledge, language, conventions, and awareness. The text that the communicator produces is only an approximation of the mental representation of what he or she wanted to communicate. Likewise, the reader interprets the text through his or her own mental representation.

How do we know that readers form mental representations and actively interpret as they read? One way is through think-aloud pro-

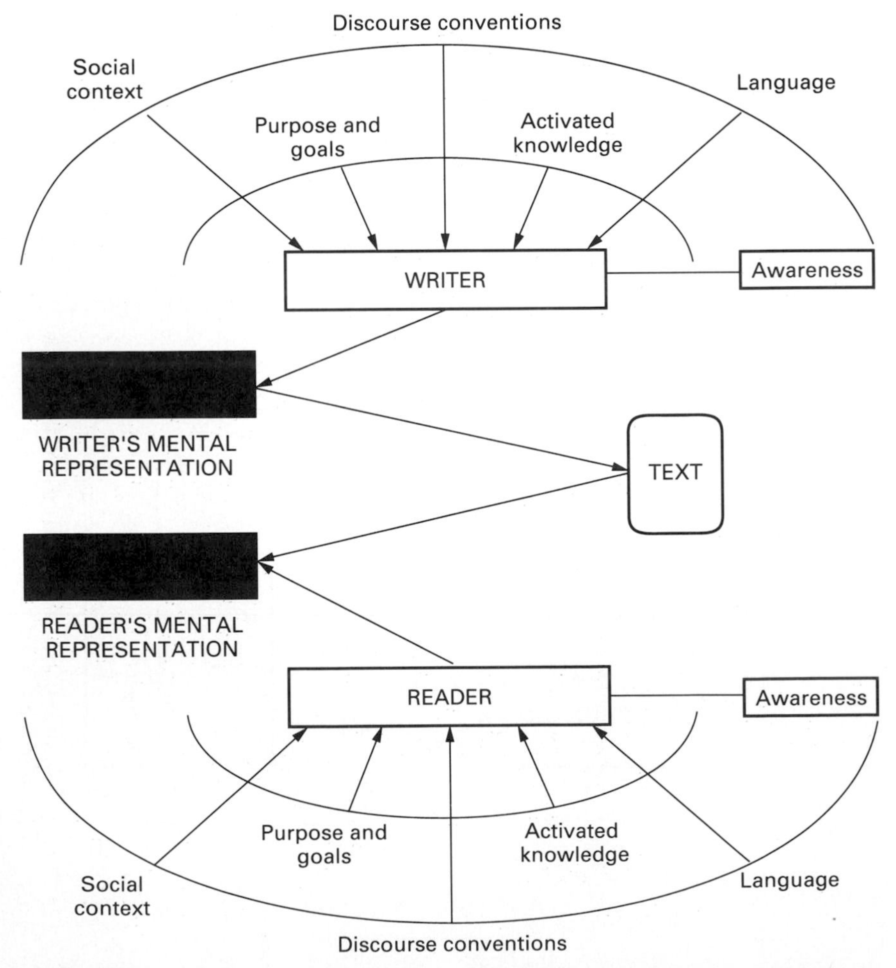

FIGURE 1.4 A Model of How Writers and Readers Interact with a Text (Flower et al., 1990)

tocols (Newell & Simon, 1972). As the name implies, people giving a think-aloud protocol say out loud whatever they are thinking as they go through a task. The *protocol* is the written record of what the person says. Think-aloud protocols have been used to understand how people solve problems and make decisions. They have also been used to understand how readers interact with documents.

Flower, Hayes, and Swarts (1983) asked people who had to use a particular federal regulation to read the regulation and think out loud. The text was typical of many bureaucratic documents: most verbs were in passive voice, the subjects of the sentences were often nouns made out of verbs, and the person or agent who should have been the subject was seldom mentioned. When Flower and her colleagues analyzed the transcripts, they found that these readers often translated the passive, nominal prose into *scenarios*. That is, the readers restated the information for themselves in terms of "who did or might do what to whom." The readers were doing a lot of work to understand the text. No wonder so few of the people who should read regulations like these actually do. Moreover, the scenarios that these readers came up with as they translated the difficult text were often wrong. The policies that the technical communicators were trying to convey were not the ones that readers thought they found in the text. The technical communicators' mental representations of the text and the readers' mental representations did not match.

Think-aloud protocols are a standard part of most usability testing and usability research. (See Chapter 11 for more information about usability testing and the place of think-aloud protocols in testing.) Using this technique, John Carroll and his colleagues at IBM's Watson Research Center discovered just how active readers and users are as they interpret documents and products. In observing and listening to secretaries who were working with a new word processing program and the manual for it, Carroll and his colleagues found that people are not passive learners. The secretaries in Carroll's studies actively interpreted the text as they tried to make sense of what happened, and they acted on the basis of their interpretations rather than on what the writer may have meant the text to say. Their interpretations came from their previous experiences in other situations and were often wrong for the program they were working with. Nonetheless, these people acted on their own explanations and on their own predictions of what would happen next. When other evidence, such as what was on the screen or what was in the manual, conflicted with their explanations or predictions, they ignored or misinterpreted the information from the screen or manual (Mack, Lewis, & Carroll, 1983; Carroll & Mack, 1984; for a general review of the work of this group, see Carroll, 1990).

In these studies, Carroll and his colleagues found that the people they observed not only interpreted as they read, but also tended to

- Resist reading
- Set their own goals
- Jump the gun to act before reading the instructions

As Carroll and Mack (1984) wrote, these people were "doing, thinking, and knowing" rather than just reading and decoding. I, and probably anyone else who has heard readers think aloud or watched people work with a product or document, have seen readers do the same with a wide variety of documents (Redish, 1988).

We cannot change the fact that readers participate actively and individually in the process of interpreting text. We should not want to even if we could because research shows us that people learn best when they are actively involved (Brown, Collins, & Duguid, 1989; Suchman, 1987). In a study of computer tutorials, Charney and Reder (1986; see also Charney, Reder, & Wells, 1988) found that readers working with a tutorial that told them exactly what to do performed only slightly better on later tasks than people who just read the tutorial without working on the computer at all. A problem-solving approach in which people had to work actively through problems as part of the tutorial resulted in faster and more accurate performance on later tasks.[1]

What You Can Do

What are the implications of the fact that readers actively interpret documents as they work with them? One implication of Flower and her colleagues' study is that the "scenario principle" is a powerful tool for writing. If the technical communicators had written that regulation in scenarios (active voice, action verb sentences with people or organi-

1. The word *active,* as in active readers, active users, active learners, is being used in two different senses in the literature at the moment, and this may be causing some confusion.

All readers are active in the sense that they use their prior knowledge and experience to understand a new document. Every reader is a thinking human being. Every reader is involved in making meaning for himself or herself by invoking and building on his or her schemata. When Carroll and others talk about readers making hypotheses about what happened, they are referring to readers being active in this sense.

Not all readers are active, however, in the sense that they want to explore new documents or products. When Mirel, Feinberg, and Allmendinger (1991) use "active learner" to refer to a certain type of learner, they are using this second sense of "active." They mean "explorer," or "risk taker," as I have used the term in the second section of the chapter. Mirel and her colleagues have observed, as we have at the American Institute for Research, that not all readers are happy exploring new products or documents.

Much of Carroll's work has been on developing materials that foster exploration. In some of his early studies, however, he found that even though his subjects did better with brief cards that forced them to learn by exploring than they did with large tomes of system-oriented documentation, some were uncomfortable with a highly exploratory approach.

zations as the subjects of the sentences), they would have increased the probability that people would actually read the text and that the readers would interpret the text to mean what the agency wanted it to mean.

Another implication is that you know how readers will interpret a document only when you try it out with readers. Think-aloud protocols can be very useful to technical communicators. Schriver (1991) has shown that, when technical communicators have access to think-aloud protocols, they produce more extensive and better revisions than they do by other methods.

READERS INTERPRET DOCUMENTS IN LIGHT OF THEIR OWN KNOWLEDGE AND EXPECTATIONS

If readers actively interpret text, what is the basis of their interpretations? The basis is their own prior knowledge and the expectations that they have about the subject matter, the type of document, and the context in which they are reading the document. Their knowledge and expectations are organized into schemata.

A *schema* (plural *schemata*) helps people to make meaning from disparate pieces of information. As Anderson and Pearson (1984) explain, schemata are networks of information connected by chronology, functions, topics, and so on. They can be fragmentary and restructured on the spot. They are multidimensional, rather like hypertext, in that one piece of information can be linked to many others, each of which has its own set of connections.

A schema is like a mental model, a way of understanding. Consider the earlier example about turning forms into letters (Bagin, 1988). More people have a schema for dealing with letters than have one for dealing with forms, and they feel more comfortable with their schema for letters.

What the Research Says

Everyone uses schemata to organize information. As they read, readers (and listeners and users) automatically access their existing schemata (Bartlett, 1932; Rumelhart, 1980). People also share elements of schemata with others in their specific cultures. For example, within Western culture, most of us have a common schema or script for "going to a restaurant" (Schank & Abelson, 1977). Suppose I said: "We went to a restaurant. The hostess showed us to a table. We sat down, and then the waiter came to give us the bill." You might say, "What? Wait a minute! What happened to the menus and the food?"

How did you know to say that? You knew because you have a schema telling you that, in a restaurant with a hostess and a waiter, people eat before paying. The words *restaurant, hostess,* and *waiter* may have made you call up that schema. If I had started with "We went to a fast-food place," you would not have expected to hear *hostess* and *waiter.* You would also probably have had a different reaction to the timing of the bill because you would have probably called up your "fast-food restaurant schema," in which paying comes before eating.

People also create and recreate their own schemata. As you read this, you are making connections in your mind to information you already know.

What You Can Do

Understanding the way that readers use their prior knowledge and expectations has profound implications for communicators organizing material both on a large scale (the document as a whole) and on a small scale (paragraphs and sentences). Let us look first at the issues that relate to whole documents and then at issues that relate to paragraphs and sentences.

Issues for Documents as a Whole. Successful documents are those that make explicit connections to readers' prior knowledge and expectations. One way to help readers is to help them activate an appropriate schema. Bransford and Johnson (1972, 1973) showed that you can do that with something as simple as a useful title or an illustration. (See Bransford & Johnson, 1972, 1973, or Duin, 1989, for a copy of the text that Bransford and Johnson used in one of their most famous experiments.) In their experiment, subjects who read the passage with a useful title recalled more than twice as many ideas as those who read the passage without a title.

Because users of a regulation come with questions in mind, a regulation that is organized by users' questions is most helpful to them (Flower et al., 1983; Redish et al., 1985). Because product users come to a manual knowing their goals or tasks, a manual that is organized by users' tasks is most helpful to them (Redish et al., 1985).

Readers should be able to glance down the table of contents or across the tabs of a document and immediately understand both the overall content and the structure of the document. The table of contents and tabs must work both as introduction and overview. Because the headings and subheadings in the document form the table of contents, technical communicators must pay special attention to how well the headings and subheadings work as a framework. If readers have only the headings and subheadings in order, without any text, will they know what your document is saying and where to find the information they need? (See Chapter 7 for more information on this subject).

Making connections to readers' schemata (prior knowledge and expectations) is critical, but so is expanding that knowledge. Otherwise, you would have no reason for writing. How do you both make connections and expand readers' knowledge?

This question is the equivalent for technical documents of what Carroll and Rosson (1987) call the "assimilation paradox." Carroll and Rosson found that, as people attempted to learn new software products, they actively interpreted what they saw on the screen and what they read in the manuals. In many cases, being able to make connections to what they already knew was helpful, but these readers also relied on their prior knowledge even when it did not apply. The paradox is that if you make no connections for readers, they will find the material very difficult to assimilate; at some point, however, that prior knowledge may hinder the same readers from moving further into the product or the document.

Fischer (1988) shows this dilemma very dramatically with the following comparison between a user's mental model of a hypothetical system and the system itself (Figure 1.5).

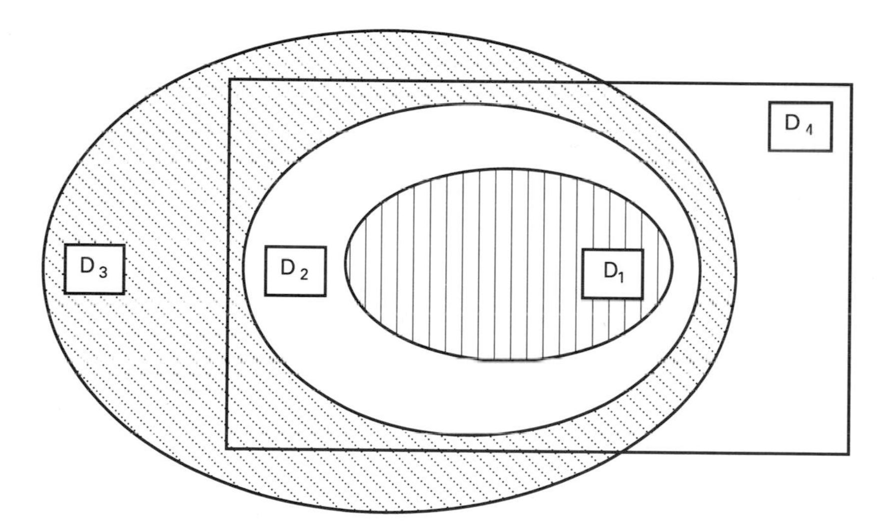

FIGURE 1.5 Comparing the User's Model of a System and the Actual System (Fischer, 1988, p. 139)

The area labeled D_1 in Figure 1.5 is the part of the system that this particular user understands and works with regularly. The larger oval, D_2, is the part of the system that the user understands less well and works with only occasionally. The rest of the oval, D_3, is the user's mental model of the system—this user's expectations of what else he or she can do in this system and how to do those other tasks. The rectangle, D_4, is the actual system. Note that the user's view, developed from an understanding of D_1 and D_2, only partially coincides with the actual system.

Discrepancies between our mental models and real systems or documents are common. Fischer drew this picture to describe the need for an intelligent help system that would

- Recognize what the user is doing or wants to do
- Evaluate how the user is trying to achieve a goal
- Construct a model of the individual user
- Decide when and how to interrupt and provide new information

With an intelligent help system like this, users could accomplish their goals more efficiently and effectively because the system would help bring the users' mental models closer to an accurate view of the system. Fischer doesn't mention it, but the opposite is also true. That is, by studying the discrepancies between users' views of the system and the actual system, product developers could change the system to be more in line with the users' models of it.

How does this relate to technical documents? Research from linguistics, rhetoric, and psychology shows us that the same principles hold for documents and products. Here are four of these principles:

- Provide an explicit schema
- Follow the given-new contract
- Maintain coherence and consistency
- Provide multiple pathways through a document

Provide an explicit schema. One way to help readers is to frame the document with a title, headings, and subheadings that show explicitly the schema you want to invoke. You can carry this further through the document by creating useful *chunks,* which are sections that you introduce by briefly elaborating an appropriate schema.

For example, Smith and Goodman (1984) compared instructions for building an electrical circuit in three formats:

- Linear (two introductory sentences and then all the instructions with no breaks or further framing).
- Structurally elaborated (that is, with brief procedural introductions to sections of the instructions). A structural (procedural) elaboration would be something like this:

  > Assembling a circuit requires that you get the major components ready, then connect them. It is often the case that the components themselves have to be assembled first.

- Functionally elaborated (that is, with brief conceptual introductions to sections of the instructions). A functional (conceptual) elaboration would be something like this:

> In a circuit, electrical current flows from a source to a consumer (i.e., to something that requires current, like a lamp). Current can flow only when the circuit's components are interconnected in a complete circle, each connection being made by a wire or other metal object that conducts electricity.

Smith and Goodman found that both types of elaboration were more useful than no elaboration at all. The structural (procedural) elaborations were somewhat more useful than the functional (conceptual) elaborations. That is, people were fastest and most accurate in completing the task—building a circuit—with the brief procedural elaborations. Brief procedural elaborations served to activate and build an appropriate schema. They also broke (chunked) the instructions into manageable pieces. Charney, Reder, and Wells (1988) also found that brief procedural elaborations in a users' guide helped readers who were not sure exactly which task they wanted to do.

Follow the given-new contract. In the 1970s, Haviland and Clark (Haviland & Clark, 1974; Clark & Haviland, 1975) showed that people who are conversing expect new information to come couched in a framework of known or previously given information. This is the *given-new contract,* and it holds for written documents as well as for conversation. Readers anticipate the given-new contract; that is, they understand new information best when it is presented in a framework of information that they already know or that they have previously been given.

For the document as a whole, the given-new contract means organizing the document with a title and headings that form a logical framework for the reader. Another way to use the given-new contract in technical documents is to set up visual patterns in the page layout that chunk information in ways that are meaningful to readers. Repeated visual patterns—the page layouts—rapidly become "given" information and help readers find what they need quickly and understand it. For example, if procedures are always given in indented, numbered lists, readers rapidly build an expectation of where to look on a page for that type of information.

Maintain coherence and consistency. Coherence in a technical document lies in the consistency of the framework that is created by the structural elements—the headings—and by the visual patterns of the pages. Creating standard visual pages can greatly facilitate reading.

In designing a report for a major company, for example, my group at the American Institutes for Research created a two-page spread, called a folio design, for each major topic. On each two-page spread, we

- Introduced the topic with the major heading, "The need for [topic]"
- Gave a specific recommendation
- Followed that with a numbered list of action steps

- Added a few short paragraphs of observations that were the rationale for the recommendation and the action steps

Busy executives who wanted to read only the recommendations could flip through the pages and find each recommendation in the same place on one page after another. If they wanted more information for a particular recommendation, they could continue down the page, grabbing the information rapidly because it was broken into small chunks and was always in the same order.

Mirel, Feinberg, and Allmendinger (1991) have recently suggested using folio designs in manuals. In their manual, each task-oriented topic gets a two-page spread. These two-page spreads incorporate the four principles of Keyes, Sykes, and Lewis (1988):

- Chunking—breaking the information into small sections that are visually distinct on the page
- Queuing—organizing the information hierarchically
- Filtering—showing the organization through levels of headings and other design features such as listing, changes in typography, and placement on the page
- Mixing modes—giving information in both words and pictures

Chapter 7 describes specific design techniques to take advantage of these principles.

Providing multiple pathways through a document. Readers bring not only their own schemata and prior knowledge to each text, but also their different cognitive, or reading, styles. Some readers are more comfortable with text, others with visuals. For many readers the synergy of text and visuals makes a document clearer. Visuals can include line drawings, charts, tables, sample computer screens, and examples set apart from the text (Redish, 1987). Chapter 6 explains how you can effectively use visuals.

Issues for Paragraphs and Sentences. People construct meaning by making sense both of the structure of the document as a whole and of the individual words, sentences, and paragraphs.

Readers' knowledge about the document as a whole influences their expectations about information they will find in that document and even colors their interpretation of individual words. Think about how you would define the word "enter" if you saw it on an income tax form, in a computer manual, or in a script for a play.

Headings can also bias expectations of the information that readers will find under them. To understand the document, however, readers must also be able to interpret the strings of words that make up the information under those headings.

Many features of the text on the level of paragraphs and sentences influence how quickly and easily readers can build meaning from the words. Chapter 5 discusses several of these features. I will discuss just two here: following the given-new contract and using the power of parallelism.

Follow the given-new contract. The given-new contract, which I discussed earlier as an issue for the document as a whole, also operates very strongly on the paragraph and sentence level (Redish, 1989). In considering scientific writing, Gopen and Swan (1990) found that "the misplacement of old and new information turns out to be the No. 1 problem in American professional writing today" (p. 555).

Dixon (1987) investigated what happens when readers get new information first and the contextual (old, given, framing) information second. Two hypotheses for how readers understand such sentences are possible:

- Readers put the new information into a mental buffer and wait for the context in order to understand it.
- Readers guess at the context and act without waiting.

The readers in Dixon's study clearly guessed. They began to act on information immediately without waiting. For example, given the following sentences, most readers drew the picture in Figure 1.6*a*.

> This will be a picture of a wine glass. Draw a triangle on top of an upside-down T.

Given the same sentences in the opposite order (action-context; new-given), however, many readers drew the "Christmas tree" in Figure 1.6*b* rather than a wine glass.

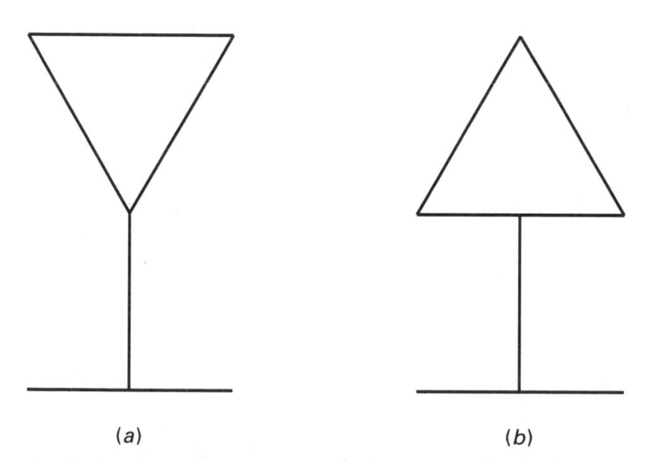

FIGURE 1.6 Different Responses for Sentences in the Order Context-Action and Action-Context: (*a*) context-action order, response correct; (*b*) action-context order, response incorrect (Dixon, 1987, p. 28)

One implication of this work is that, even on the sentence level, writers should follow the order in which readers will expect the material or in which users will do their tasks. Consider the following examples:

Poor order:

Enter ABC at the A> prompt to install ABC.

Better order:

To install ABC:

1. At the A> prompt, type: ABC
2. Press Enter.

Readers of the first example have to wait until the end of the sentence to be sure that the sentence relates to what they want to do (install ABC) and that they are in the right place (at the A> prompt). Readers who guess and start to type before finishing the sentence may find themselves in trouble if they are not at the A> prompt or if they don't want to install ABC. The first, poor, example gives readers instructions in the reverse order from the way they must act. They must first verify that they want to install ABC, then they must see that they are at the A> prompt or get there, then they must type the appropriate letters, and only last do they actually "Enter" anything, that is, press the Enter key.

Structural principles such as the given-new convention are far more important in developing useful, understandable documents than are readability formulas. Readability formulas are based on the outdated, static, linear model of Figure 1.2 (on p. 22), which assumes that the same meaning exists in the text for all readers (Redish & Selzer, 1985; Duffy, 1985). Most readability formulas also rely on only one sentence feature—length—and one semantic feature—word length or word frequency.

Duffy and Kabance (1982) found that shortening sentences and simplifying words in several reading passages led to improved scores on readability formulas but did not make the passages easier to understand. Only in one case did the Navy recruits in their study do better on a comprehension test with the revised passages than with the original ones.

Olsen and Johnson (1989) analyzed Duffy and Kabance's revised passages in an attempt to understand these results. They found that, although Duffy and Kabance's revised passages had better scores on readability formulas, they were less coherent and cohesive in their given-new links. In breaking up the sentences, Duffy and Kabance had moved information into positions that violated the given-new contract.

Olsen and Johnson suggest that Duffy and Kabance's revisions did not improve comprehension because they did not maintain the given-new contract. Shortening sentences while disrupting the given-new contract may improve the score on a readability formula but hinder real readability—that is, people trying to understand the meaning of a text, as measured by a comprehension test.

Chapter 2 discusses this point further. Chapter 5 presents specific writing techniques for attaching new information to existing knowledge.

Use the power of parallelism. A second important principle in meeting readers' needs at the paragraph and sentence level is to structure the text so that it places the least possible burden on the reader. One way to reduce the reader's burden is to create and use recurring patterns in the text, to use parallel syntactic structures.

Readers respond to recurring patterns of text. They also try to impose patterns on text. If you use visual techniques such as bulleted lists for showing off the parallelism, the parallel text is also much easier to grasp quickly.

For example, compare the two versions of the same passage in Figures 1.7 and 1.8. Figure 1.7 is the original text. It comes from an insurance policy that meets the readability standards of all the states that use a readability formula as the measure of "plain English." How "plain" do you think it is?

Option Two—Premium Reduction

Dividends will be used to pay or reduce any premium then due provided any amount of that premium remaining is paid by the end of the grace period. If the dividend is larger than the premium due, any excess will be paid in cash to the owner.

FIGURE 1.7 The Original—without Parallelism

Readers trying to understand this paragraph have to set up the parallelism for themselves. (They also have to translate the passive, nominal prose into personal, active sentences, that is, into scenarios—see the discussion of readers translating text into scenarios earlier in this chapter. See also Chapter 9, for more information on using passive and active verbs.)

When the technical communicator makes the parallelism in the choices obvious, the paragraph puts much less burden on the reader and becomes much easier to understand quickly and correctly. Figure 1.8 is my revision.

Option Two—Using Dividends to Help Pay the Premium

If you choose this option, we will use each dividend to pay all or part of the next premium that is due.

- If the dividend is less than the premium, you must pay the difference before the end of the grace period.
- If the dividend is more than the premium, we will send you a check for the extra amount.

FIGURE 1.8 A Revision–with Parallelism

Technical communicators can help readers find and understand information in their documents by using techniques like the ones that I have described in this section. For the document as a whole, these techniques are as follows:

- Provide an explicit schema through a useful title and table of contents.
- Follow the given-new contract; for example, use headings that mirror readers' prior knowledge and expectations.
- Maintain coherence and consistency; for example, use repetitive, consistent page layouts.
- Provide several pathways through a document; for example, give information visually as well as in prose.

The techniques for paragraphs and sentences are as follows:

- Follow the given-new contract on the local as well as the global level; for example, give the context before the instruction in a manual.
- Make parallel concepts clear by writing parallel sentences.

These six are only a few of the techniques that work well for making documents easy for busy readers who are trying to find and understand information quickly. They are a base for you to build on as you read the other chapters in this book.

SUMMARY

Readers are not passive vessels into which technical communicators can pour information. Readers are active participants in the communication process. They decide whether to pay attention to a document and how much attention to pay. Technical documents, in the workplace or the home, are usually meant for busy people who need to accomplish a goal or task that goes beyond just reading the document. Because the document is a tool to help accomplish the goal, readers want to find what they need in the document quickly and understand it easily.

Readers actively interpret what they read, calling on *schemata* or mental models that they have developed from prior knowledge and experiences. Technical communicators can, however, improve the chances that readers will understand their messages by taking into account readers' knowledge, expectations, and styles and by using techniques that have been shown to match the way that readers approach documents.

ACKNOWLEDGMENTS

This chapter is based on workshops developed with Ann Hill Duin of the University of Minnesota for the International Technical Communication Conferences in 1989 and 1990 and on talks that I was invited to give at the International Technical Communication Conference and to Hewlett-Packard's Learning Products Council in 1990.

LIST OF TERMS

chunking Breaking the text into short, meaningful pieces. Frequent, useful headings and subheadings support chunking. Short paragraphs and lists support chunking.

given-new contract Says that listeners and readers expect communicators to use information they already know as the context for new information. Listeners and readers understand most easily when technical communicators link new information to information that readers bring to the document or that technical communicators have already given in the document.

reading to do Reading documents to "do" or act. If readers need the documents for reference later, they can go back and look at them again. Many documents that people in business deal with are used for reading to do.

reading to learn Reading to learn information that the reader will have to remember later. Textbooks are used for reading to learn. Reading-to-do and reading-to-learn materials require different communication techniques.

reading to learn to do Describes the reading of certain documents, such as computer tutorials and users' guides, that serve an intermediate purpose. Readers go to them wanting to accomplish their own tasks quickly. They do not want to spend time reading. However, they also want to learn how to do tasks so they will not have to look them up each time. Technical communicators can learn techniques to facilitate reading to learn to do.

satisficing A term borrowed from the work of the noted economist Herbert Simon. People trade off the time and effort it takes to learn more or to find a better solution against accepting a less than optimal way of doing something or a less than perfect solution to a problem.

scenario principle Writing in the active voice with action verbs and the doer of the action as the subject of the sentence. Flower and her colleagues (1983) found that readers trying to make sense of bureaucratic prose translated the text into scenarios.

schema (pl. schemata) Structures that people form by developing associations among separate pieces of information. A schema (plural, schemata) helps people to make meaning of disparate pieces of information.

think-aloud protocol Used to describe the process in which a reader or user says all of his or her thoughts out loud while reading a document or doing a task. The entire session is tape-recorded, and the tape can be transcribed and analyzed. Think-aloud protocols are very useful for gaining insights into how people solve problems and understand documents.

REFERENCES

Anderson, R. C., & Pearson, P. D., (1984). A schema-theoretic view of basic processes in reading comprehension. In P. D. Pearson (Ed.), *Handbook of reading research*, (pp. 255–291). New York: Longman.

Bagin, C. B. (1988, January). Sometimes letters are better than forms. *Simply Stated . . . in Business, 20,* 1–2.

Bagin, C. B., & Rose, A. M. (1991a, February–March). Worst forms unearthed: Deciphering bureaucratic gobbledygook. *Modern Maturity,* pp. 64–66.

Bagin, C. B., & Rose, A. M., (1991b). *Frustrated and confused: What people told us about bad forms,* Washington, DC: American Institutes for Research. [a more detailed report on the AARP study of 1991a]

Bartlett, F. C. (1932). *Remembering.* Cambridge MA: Cambridge University Press.

Bransford, J. D., & Johnson, M. K. (1972). Contextual prerequisites for understanding: Some investigations of comprehension and recall. *Journal of Verbal Learning and Verbal Behavior, 11,* 717–726.

Bransford, J. D., & Johnson, M. K. (1973). Considerations of some problems of comprehension. In W. Chase (Ed.), *Visual information processing.* New York: Academic Press.

Brown, J. S., Collins, A., & Duguid, P. (1989, January-February). Situated cognition and the culture of learning. *Educational Researcher,* pp. 32–43.

Carroll, J. M. (1990). *The Nürnberg Funnel.* Cambridge MA: MIT Press.

Carroll, J. M., & Mack, R. L. (1984). Learning to use a word processor: By doing, by thinking, and by knowing. In J. C. Thomas & M. L. Schneider, (Eds.), *Human factors in computer systems* (pp. 13–51). Norwood, NJ: Ablex. Reprinted in R. M. Baecker & W. A. S. Buxton (Eds.), *Readings in human computer interaction* (pp. 278–297). San Mateo, CA: Morgan Kaufmann, 1987.

Carroll, J. M., & Rosson, M. B. (1987). "The paradox of the active user. In J. M. Carroll (Ed.), *Interfacing thought: Cognitive aspects of human-computer interaction* (pp. 80–111). Cambridge, MA: MIT Press.

Charney, D. H., & Reder, L. M. (1986). Designing interactive tutorials for computer users: Effects of the form and spacing of practice on skill learning. *Human-Computer Interaction, 2,* 297–317.

Charney, D. H., Reder, L. M., & Wells, G. W. (1988). Studies of elaboration in instructional texts. In S. Doheny-Farina (Ed.), *Effective documentation: What we have learned from research* (pp. 47–72). Cambridge, MA: MIT Press.

Clark, H., & Haviland, S. (1975). Comprehension and the given-new contract. In R. Freedle (Ed.), *Discourse production and comprehension.* Hillsdale, NJ: Lawrence Erlbaum.

Dixon, P. (1987). The processing of organizational and component step information in written directions. *Journal of Memory and Language, 26,* 24–35.

Duffy, T. M. (1985). Readability formulas: What's the use? In T. M. Duffy & R. M. Waller (Eds.), *Designing usable texts* (pp. 113–143). New York: Academic Press.

Duffy, T. M., & Kabance, P. (1982). Testing a readable writing approach to text revision. *Journal of Educational Psychology, 74,* 733–748.

Duin, A. H. (1989). "Factors that influence how readers learn from text: Guidelines for structuring technical documents. *Technical Communication, 36,* 97–101.

Fischer, G. (1988). Enhancing incremental learning processes with knowledge-based systems. In H. Mandl & A. Lesgold (Eds.), *Learning issues for intelligent tutoring systems* (pp. 138–163). New York: Springer-Verlag.

Flower, L., Hayes, J. R., & Swarts, H. (1983). The scenario principle. In P. V. Anderson, R. J. Brockmann, & C. R. Miller (Eds.), *New essays in technical and scientific communications: Research, theory, and practice* (pp. 41–58). Farmingdale, NY: Baywood.

Flower, L., Stein, V., Ackerman, J., Kantz, M. J., McCormick, K., & Peck, W. C. (1990). *Reading-to-write: Exploring a cognitive and social process.* New York: Oxford University Press.

Gopen, G. D., & Swan, J. A. (1990). The science of scientific writing. *American Scientist, 78,* 550–558.

Haviland, S., & Clark, H. (1974). What's new? Acquiring new information as a process in comprehension. *Journal of Verbal Learning and Verbal Behavior, 13,* 512–521.

Holland, V. M., & Redish, J. C. (1982). Strategies for using forms and other public documents. In D. Tannen (Ed.), *Proceedings of the Georgetown University Round Table on Languages and Linguistics: Text and Talk* (pp. 205–215). Washington, DC: Georgetown University Press.

Keyes, E., Sykes, D., & Lewis, E. (1988). Technology + design + research = information design. In E. Barrett (Ed.), *Text, ConText, and HyperText: Writing with and for the computer* (pp. 251–264). Cambridge, MA: MIT Press.

Mack, R. L., Lewis, C. H., & Carroll, J. M. (1983). Learning to use word processors: Problems and prospects. *ACM Transactions on Office Information Systems 1*, 254–271; reprinted in R. M. Baecker & W.A.S. Buxton (Eds.), *Readings in human-computer interaction* (pp. 269–277). San Mateo, CA: Morgan Kaufmann, 1987.

Mikulecky, L. (1981). *Job literacy: The relationship between school preparation and workplace actuality* (Report to the Department of Education), Bloomington, IN: Indiana University.

Mirel, B., Feinberg, S., & Allmendinger, L. (1991). Designing manuals for active learning styles. *Technical Communication*, First Quarter, 75–87.

Newell, A., & Simon, H. (1972). *Human problem solving*. Englewood Cliffs, NJ: Prentice-Hall.

Norman, D. A. (1988). *The design of everyday things*. New York: Basic Books. [originally titled *The psychology of everyday things*]

Olsen, L. A. & Johnson, R. (1989). Towards a better measure of readability: Explanation of empirical performance results. *Proceedings of the 36th International Technical Communications Conference*. Chicago, pp. RT 7–10.

PC Magazine. (1988, March 29). *PC Magazine* survey, pp. 40–41.

Redish, J. C. (1987). Integrating art and text. *Proceedings of the 34th International Technical Communications Conference*. Denver, pp. VC 4–7.

Redish, J. C. (1988). Reading to learn to do. *Technical Writing Teacher, xv*, 223–233; reprinted in *IEEE Transactions on Professional Communication, 32* (December 1989), 289–293.

Redish, J. C. (1989). Writing in organizations. In M. Kogen (Ed.), *Writing in the business professions* (pp. 97–124). Urbana, IL: National Council of Teachers of English.

Redish, J. C., Battison, R. M., & Gold, E. S. (1985). Making information accessible to readers. In L. Odell & D. G. Goswami (Eds.), *Writing in nonacademic settings* (pp. 129–153). New York: Guilford Press.

Redish, J. C., & Selzer, J. (1985). The place of readability formulas in technical writing. *Technical Communication, 32*, 46–52.

Rose, A.M. (1981). Problems in public documents. *Information Design Journal, 2*, 179–196.

Rumelhart, D. E. (1980). Schemata: The building blocks of cognition. In R. J. Spiro, B. C. Bruce, & W. C. Brewer (Eds.), *Theoretical issues in reading comprehension* (pp. 33–58). Hillsdale, NJ: Lawrence Erlbaum.

Schank, R. C., & Abelson, R. P. (1977). *Scripts, plans, goals, and understanding*. Hillsdale, NJ: Lawrence Erlbaum.

Schriver, K. A. (1991). Plain language through protocol-aided revision. In E. R. Steinberg (Ed.), *Plain language: Principles and practice* (pp. 148–172). Detroit: Wayne State University Press.

Simon, H. (1976). *Administrative behavior* (3rd ed.). New York: Macmillan.

Smith, E. E., & Goodman, L. (1984). Understanding written instructions: The role of an explanatory schema. *Cognition and Instruction, 1,* 359–396.

Sticht, T. (1985). Understanding readers and their uses of texts. In T. M. Duffy & R. Waller (Eds.), *Designing usable texts* (pp. 315–340). Orlando, FL: Academic Press.

Suchman, L. (1987). *Plans and situated actions.* New York: Cambridge University Press.

Sullivan, P., & Flower, L. (1986). How do users read computer manuals? Some protocol contributions to writers' knowledge. In B. T. Petersen (Ed.), *Convergences: Transactions in reading and writing* (pp. 163–178). Urbana, IL: National Council of Teachers of English.

Wright, P. (1981). The instructions clearly state . . . Can't people read? *Applied Ergonomics, 12,* 131–142.

Wright, P. (1987). Writing technical information. In E. Z. Rothkopf (Ed.), *Review of research in education* (Vol. 14, pp. 327–385). Washington, DC: American Educational Research Association.

Wright, P. (1988). Issues of content and presentation in document design. In M. Helander (Ed.), *Handbook of human-computer interaction,* (pp. 629–652). New York: North-Holland.

Wright, P. (1989). Can research assist technical communicators? *Proceedings of the 36th International Technical Communications Conference.* Chicago, pp. RT 3–6.

Wright, P., Creighton, P., & Threlfall, S. M. (1982). Some factors determining when instructions will be read. *Ergonomics, 25,* 225–237.

CHAPTER 2

THE DESIGN DRAFT–ORGANIZING INFORMATION

Heather Fawcett, Samuel Ferdinand, and Ann Rockley

Information Design Solutions

INTRODUCTION

With a firm understanding of your readers, you are ready to embark on your project. This chapter explains how to write a *design draft*. A design draft is your first attempt at organizing your information. The design draft includes, at the least, a list of topics in the order in which you plan to present them. Your design draft can take the form of a traditional outline, a visual representation (called a *mindmap*), or a detailed design of each page, screen, or unit of information (called a *storyboard*).

The design draft provides editors, managers, other writers, product developers, market planners, usability consultants, and others with a detailed representation of how you plan to organize your information. They can get an idea of the scope of the project from the design draft. The technical communication department and other departments can also benefit from a detailed design draft. For the technical communication department, a design draft

- Clarifies the communicator's purpose
- Provides an outline for writing
- Helps communicators work together in a team by ensuring that all have the same understanding of the project
- Allows topics to be split among writers in the team
- Provides an estimating and scheduling tool for the technical communications department manager

For other departments, a design draft

- Verifies that the documentation effort is in line with the product's design and purpose
- Provides a way to coordinate the writing effort with other departments' efforts
- Provides a basis for review and discussion

This chapter discusses the theory and techniques for developing a good design draft. Specifically, this chapter describes

- The benefits of effective organization
- A five-step process for developing a design draft:
    1. Determine the purpose of the information by
        a. Determining what you're writing about
        b. Determining who you're writing for
        c. Determining the objectives of the information

2. Select the medium of communication (such as print or video)
3. Organize the information by
   a. Selecting appropriate organizational patterns
   b. Organizing the information using one or more organizational techniques
4. Formalize the organization
5. Evaluate the organization

BENEFITS OF EFFECTIVE ORGANIZATION

Good organization doesn't just happen as you write. As an architect creates a plan or blueprint of a building before starting construction, so a technical communicator creates a plan of the organization—a design draft—before writing. Before creating the design draft, however, you must understand what organization is and why it is important to readers.

What Is Organization?

The process of organizing information is giving it order, making it comprehensible to readers. This involves generating ideas, imposing structure on ideas, exploring relationships between ideas, and arranging and evaluating these ideas.

The organization of information is shaped by its readers, content, purpose, and communication medium. The following summarizes the influence of each factor on the organization of information:

Readers — Readers' needs often influence the ordering and *chunking*—i.e., breakdown—of information. For example, readers inexperienced in a particular subject will require that information be broken down to its most basic level. They need conceptual information up front where it won't be missed, while the more knowledgeable readers need it in an appendix where it can be referred to only if required.

Purpose — Documents that serve different purposes are also organized differently. A document that is designed to instruct readers on the use of something will be organized differently from a document that is designed to be used for reference. For example, instructional information is often organized in procedures that readers can follow in sequence; reference information is often organized alphabetically so readers can quickly look up topics.

Content — The subject matter influences the organization of the information. Complex subjects are often also complex in structure. For example, procedures for operating a nuclear power plant will have a deep structure with many levels of subsections and interconnections, while procedures for assembling a lawn mower will have a broader structure with few subsections and interconnections.

Media — The choice of media can influence the organization of information. An online document may have to be written in small chunks or in expanding levels of detail, depending on the software and hardware. For example, the software or the small size of the screen may limit the amount of information that can appear on the screen.

Why Good Organization Is Important

Good organization helps readers process information efficiently, by taking into account how the brain processes information. Good organization also presents information in such a way that people process it with the least effort and with the most understanding.

As noted in the last chapter, when people are confronted with some new information they try to make sense of it by relating it to something they already know, a schema. A schema may be a concept, an object, an abstract idea, an action, a setting, or even an event. An example of an event used as a schema is someone telling a story and the listener remembering something similar that happened to him or her. Because the listener already has a mental representation of a similar event, the connection is made between the two.

Schemata help people make sense out of the world. They give people a framework for understanding new information. Besides aiding learning and recall, schemata help people generate expectations and make inferences (Kent, 1987). For example, when people encounter a computer, they expect to encounter at least some of the elements associated with the schemata of "computer," such as a keyboard, monitor, and mouse. Conversely, when people encounter one or more elements of the concept "computer," they infer or expect that the object is a computer.

Schemata also help people make inferences when they read. Readers use background knowledge or schemata to supply the underlying information. Instead of processing facts in a paragraph one by one, readers get a general impression or "gist" of the information. The inferences or gists are what readers remember, not the details (Flower, 1989).

How to Organize for Learning and Recall

If you can elicit an appropriate schema for readers, the new information will be more easily learned and remembered. To help you do so, use the following guidelines when planning and writing.

Relate New Information to What Readers Already Know. Provide readers with familiar information before introducing new material. Figure 2.1 shows an example from a user guide for students that introduces the concept of a computer notebook and editor. It starts with a description of the traditional school notebook that students are familiar with, moves on to relate this to a computer notebook, and finally explains the role of the editor in creating a computer notebook.

LESSON 1: CREATING NOTEBOOKS

What is a Notebook on a Computer?

At school you write your lessons or homework in a *notebook*. Your teacher may write down the names of students who are away from class in a notebook. Your parents may have a notebook where they write down their shopping list or important phone numbers. Anything you can write down in a regular notebook, you can type into a notebook on a computer. You can have many different notebooks, with different things in them.

What is the Editor?

The *Editor* is a place where you can make a notebook on the computer. Once inside the Editor, you can begin typing into your notebook.

The Editor also lets you make changes to a notebook. In fact, the word *edit* means to make changes to a notebook. The Editor is quick and easy to learn, since you use little pictures, called *icons,* to tell it what to do. For instance, there is a pair of scissors you use to cut words out of your notebook, and a pot of paste to paste words into your notebook.

FIGURE 2.1 Relating New Information to Known (Copyright © 1987, Unisys Corporation)

Help Readers Draw the Right Inferences from the Material. Readers process information more efficiently if they can quickly identify the type of information being presented. Give readers the organization they expect. For example, college students expect to find a summary of each chapter in a textbook (notice how this book follows the convention).

Provide Readers with the "Gist" of the Information. Give readers a framework for understanding the information. When organizing, present general information first so that the appropriate schema is elicited; this also lets readers make sense of the specific information presented later. For example, instructions for assembling a complicated piece of machinery should present an overview of the assembly process before detailing the steps in the procedure. The first sentence of this paragraph provides you with the "gist" of the paragraph.

Limit the Amount of New Information You Present at a Time. Don't overtax readers. Build on previous material.

- When organizing information, start with general information and provide detail in layers.
- Present features one at a time to avoid overburdening readers.
- Present information in small chunks that readers can easily comprehend. For example, break down a complicated procedure into steps to make the task more manageable for readers rather than covering the entire procedure in a single, long paragraph. The first example below shows a procedure presented in one paragraph; immediately following is the same procedure broken down into numbered steps and small paragraphs.

Rub some flour on your hands. Fold the dough on itself. Using the heel of your hands, push the folded dough away from you with a rolling motion. Give the dough a one quarter-turn and repeat the "fold and push" operation. Continue until the dough is smooth and elastic (8 to 10 minutes). Well-kneaded dough looks satiny and has tiny gas bubbles just below the surface. If the dough becomes sticky, sprinkle the board with flour under the dough and rub more flour on your hands.

1. Rub some flour on your hands.
2. Fold the dough on itself.
3. Using the heel of your hands, push the folded dough away from you with a rolling motion.
4. Give the dough a one quarter-turn and repeat the "fold and push" operation.
5. Continue until the dough is smooth and elastic (8 to 10 minutes). Well-kneaded dough looks satiny and has tiny gas bubbles just below the surface.

If the dough becomes sticky, sprinkle the board with flour under the dough and rub more flour on your hands.

DEVELOPING THE DESIGN DRAFT

Determine the Purpose of the Information

Before you begin organizing the information and determining the communication medium, you need to determine why you're writing. What is the purpose of the information? To determine the purpose, you must determine the following:

1. What you are writing about (the subject)
2. Who you are writing for (the readers)
3. What you want the readers to achieve after reading the information (the behavioral objectives)

Determine What You Are Writing About. First you must know what subject you're writing about. Sometimes, you're told the subject. For example, someone might assign you the project of writing the procedures for resolving problems with a laser printer. Other times, you determine the subject yourself. You might be proposing an article for a scientific journal. Because you work in a biotechnology company, you might choose to write about an application of recombinant DNA.

Whether you are told what you're writing about or choose the subject yourself, you need to learn enough about it to make informed decisions about communicating the information. You can learn about your subject by reading about it, talking to technical experts, and, in the case of product documentation, using the product.

Following are some of the questions you might ask for various types of documents:

Type of Document	Questions to Ask	Example
Product manual	What is the purpose of the product? Who will use the product? Why will they use the product? How will they use the product? Where will they use the product? What is its competition? What are the competitors' strengths? Weaknesses? (Understanding competition is essential to ensuring the success of your product.)	Suppose you are writing the user's guide for a new cooking range. You might ask what features the range offers (such as a built-in grill), what types of homes this range is intended for (small apartments, luxury homes), and what its competition is (ranges that cost twice as much but have no additional features).

(continued)

Type of Document	Questions to Ask	Example
Technical report	What aspect of the subject area are you writing about? How does this information fit into the larger field (for example, how does recombinant DNA fit into the field of biology)? What is the theoretical focus of the research? How does this compare to the theoretical focus of other work in the field (is it mainstream or radical)? What are the practical applications of this research?	Suppose you are writing an article on the current status of superconductors. You might compare recently reported research and ask what relevance each has for commercial applications of this technology. You might also mention which commercial applications of superconductors are expected and when they might be available.
Sales proposal	Who is the customer? What are the customers' requirements (these are usually documented in a publication called a Request for Proposal)? What products and services do you offer that meet these requirements or can be modified to meet them (if necessary)? Which other organizations are submitting proposals (if this information is available)? What products and services might they propose to meet the customers' requirements? How do your products compare?	Suppose you are writing a proposal to sell cleaning services for a large office complex. You might determine when the customer wants the complex cleaned (during work hours, after hours?), cleaning services needed (emptying trash every day, vacuuming twice a week), and which services you offer (you can offer recycling, which the customer wants but is not required). You might also determine who is competing for the contract.

Determine Who You're Writing For. The first part of this book focuses almost exclusively on readers. They're central to any communication effort. Not surprisingly, you begin planning a communication project by creating a profile—or mental picture—of your readers. To organize information so it is appropriate to your *particular* readers, you need to understand them in depth. Specifically, you need the answers to the following questions:

- Who are the readers?
- What do they already know?
- How are they going to use the information?
- What are the questions they will ask?

Who are the readers? First, get some general information about your readers. How old are they? What is their job description? What is their average level of education? What are their day-to-day work activities? This information gives you a general insight into your readers and their information needs.

Recognize, however, that your readers are more than generalities. They're individuals like Sheri, a 40-year-old, college-educated executive secretary who not only types memos, but usually writes them for her boss, or Jed, a 32-year-old application programmer who coordinates programming projects rather than writing programs. If possible, spend some time with your intended readers so you're writing to people that exist in the real world.

What do readers already know? Before reading, people usually know something about the subject. Determine what readers already know. You can get this information from the reader directly through interviews, questionnaires, surveys, and focus groups (extended interviews). Indirect sources include people in close contact with the reader; in companies, these include marketing and technical support personnel. Once you understand the readers' knowledge base you can determine how best to build upon it. For example, suppose you are planning a document on basic word processing techniques. Through interviews with your potential readers, you find out that most are new at producing documents on the computer but are adept at producing them on an advanced typewriter. Knowing this, you can plan your document to use comparisons between the two forms of word processing, thus building upon your readers' knowledge.

How are readers going to use the information? As mentioned in the first chapter, readers read information in at least three ways (Diehl & Mikulecky, 1981):

- *Reading to do:* reading to perform a task. When doing this, readers want to be guided through the process. For example, you've just bought a swing set for your children that needs to be assembled. You read the assembly instructions to find out how to do the task. The instructions take you through each step of the process.
- *Reading to learn:* reading to learn about something. When doing this, readers want the general information first, then specific information, as they gradually pick up more complex concepts and information. For example, you've decided to learn more about solar heating. You first require some general information about solar energy (a general definition and other facts about the development of the field) and then more specific information about how solar heating works, how efficient it is compared to other heating sources, and some specific examples of how solar heating is used in homes or businesses.

- *Reading to assess:* reading to assess the document's contents. When doing this, readers are often figuring out what information is available and storing a mental reference to the material for later use. People tend to remember information in the visual form in which it appears, so readers who are reading to assess rely heavily on the visual organizational cues such as headings and graphics. For example, you've just installed a VCR using the manual. You skim the rest of the manual to assess whether it may be useful to you (does it contain warranty information, troubleshooting information, a short list of VCR settings?). Later, when you want to use an obscure feature, you remember you saw a chart on a left-hand page of the manual with a list of all the VCR settings.

Chapter 1 suggests a fourth reason that readers read:

- *Reading to learn to do:* reading to learn how to perform a task. Users sitting down to a computer tutorial are both reading to learn and reading to do. They need conceptual information about the program as well as procedural information to help perform the tasks. Most important, users need information that will help them transfer what they learn in the tutorial to real tasks. Examples and simulations help users make this transfer.

What questions will readers ask? Put yourself in your readers' place and ask the questions they would ask. These questions can help determine what information to include, how to organize the information, and the communication medium to use. To make this procedure easier, try asking "Who?", "What?", "Where?", "When?", "Why?", and "How?" For example, in assembling a swing set, readers might first ask the following questions:

- What tools do I need?
- Where should it be placed (grass, sand, acceptable slope)?
- Who can use it (weight and height restrictions)
- How do I assemble it?

These suggest readers need answers to specific questions ("What tools do I need?", "Where should it be placed?", "Who can use it") as well as procedural information ("How do I assemble it?", "How do I maintain it?"). The first two questions at least must be answered before the questions on assembly. Given the number and complexity of questions, a small brochure or sheet of instructions would be an effective communication medium. (See "Example of the Design Draft Process" later in this chapter for a detailed example of a reader profile.)

Define the Objectives of the Document. Once you have determined what you're writing about and who you're writing for, you need to determine what readers should be able to do *after* reading the information. This, in turn, suggests what information to include, how to organize it, and in what medium to present it.

How to write objectives. You state what readers should be able to do after reading your information in statements called *terminal objectives.* Besides helping you organize information, terminal objectives help you test the effectiveness of the information. You do so by observing whether people can correctly achieve what the terminal objective states. Chapter 11 explains how to test the effectiveness of information.

A terminal objective looks like this:

> With the assistance of the manual, readers should be able to install the VCR within 30 minutes without any errors.

According to Dick and Carey (1990) a terminal objective has three parts: observable action, level of quality, and conditions. Definitions and examples of each part are shown below.

Part	Definition	Example
Observable action	Description of behavior or skill the learner will demonstrate	"install the VCR"
Level of quality	Level of quality to be achieved	"within 30 minutes" "without any errors"
Conditions	Conditions under which the action is performed; that is, tools or references that are available while performing tasks	"with the assistance of the manual"

Sometimes, you want readers to acquire knowledge—not perform tasks—as a result of reading information. Unfortunately, knowledge cannot be directly observed. For example, you cannot observe someone "knowing" superconductivity, but you can observe someone defining the term and stating the basic principles underlying superconductivity. Thus, you should express knowledge objectives in terms of actions that you can observe, such as defining, explaining, and demonstrating.

Incorrect Example: Understand what a mouse does.
Correct Example: Define the term *mouse.*

Terminal objectives state broad skills and knowledge readers should have acquired after reading information. Yet, readers often need to know how to perform other skills and need background information to successfully achieve the terminal objectives. These prerequisite skills and knowledge are called *enabling objectives.* You write enabling objectives the same way you write terminal objectives.

For example:

Terminal Objective: Install the VCR
Enabling Objectives: Connect cable TV cables to VCR.
Connect TV to VCR.

Enabling objectives are the information needed to successfully master the terminal objectives. Thus, when you write enabling objectives, you are also identifying information to include in your document. You write enough enabling objectives to fill the gap in knowledge or skills between what readers already know and what they need to know to successfully master the objective.

How to perform a task analysis. You determine terminal and enabling objectives by performing a *task analysis*. A task analysis identifies all the tasks performed by the intended reader. When there are multiple readers, or audiences, a task/audience matrix shows which tasks need to be performed by each audience group. The example task/audience analysis below shows two audience groups for a telephone system: telephone operators and telephone technicians. Notice that the telephone technicians need to perform tasks pertaining to both the operation and repair of the telephone system while the telephone operators need to perform only the operational tasks.

Task	**Telephone Operator**	**Telephone Technician**
Forwarding a call	√	√
Putting a call on hold	√	√
Redialing a number	√	√
Parking a call	√	√
Transferring a call	√	√
Programming phone numbers	√	√
Taking apart the receiver		√
Taking apart the handset		√
Checking the phone connection		√
Checking the phone line		√

Each of these tasks can be broken down into subtasks. For example, "redialing a number" contains the subtasks: "dialing the original number" and "pressing the redial key." These become the enabling objectives for "redialing a number." Keep breaking down the tasks until the gap disappears between what readers already know and what they need to know to perform the task successfully. For example, to "press the redial key," the reader must be able to "recognize the redial key" on the handset.

What you do with objectives. Terminal objectives are often represented in printed or online documents as headings, and the enabling objectives as subheadings or steps in the procedure. For example, the telephone system manual may have a chapter called "Operating Your Telephone System" which contains a section called "Redialing a Number." This section, in turn, contains the steps "Making a Call" and "Pressing the Redial Key." Setting the terminal and enabling objectives also suggests which information you should include in a document. Include information that helps readers achieve the objective. For example, the enabling objective "recognize the redial key" could be achieved with a picture of the phone keypad with the redial key highlighted.

Choose a Communication Medium

Once you've set the objectives for your information, you need to select a suitable communication medium. A communication medium is a tool for publishing information, such as printing information or presenting it online. Following are the most commonly available media and a description of their strengths and weaknesses.

Medium	Strengths	Weaknesses	Relative Cost to Develop
Print	Portable. Compact. Flexible presentation (page layout, fonts, graphics). Readers can modify (with markers or comments on pages).	Not instantly available. Not interactive. Often used as last resort by readers.	Inexpensive to moderately expensive.
Video	Can show exactly what readers should do or what they will see. Familiar medium.	Requires specific equipment to view. Very difficult to update.	Expensive.
Online	Context sensitive. Interactive. Easy to update. Readers can select their route through information. Can include other media such as audio and video.	CRT resolution and size can reduce readability. Loss of traditional navigational aids like page numbers, dividers. Users can become lost (especially when following hypertext links). Scope of information may be unclear.	Moderately expensive to expensive if audiovisual material is included.

The medium you select depends on many factors, some of which are shown below.

Factor Influencing Medium	Example
Content of information	Pictures of detailed and complicated processes or machinery reproduce best in print or on video.
Purpose of information	Reading-to-learn information is often best presented in print or video where reading is easier; online is better suited to reading-to-do information when the tasks being documented are performed on a computer.
Stability of information	Rapidly changing information, such as price lists, may be best presented online where the information can be quickly updated and transferred between sites.
Reader profile	Readers who are computer experts will be more comfortable with online documents than novice computer users.
Place of reading	Textbooks for students are normally printed because most of their reading is done at home where they may not have access to a computer or a video recorder.
Cost of development	Video is too expensive for most small communication projects; many documents can be produced inexpensively in print.
Resources available	Online instructional documents require several types of expertise to be done well, including instructional design, technical communication, and programming.

Consider, for example, how to select a medium to explain safety procedures on a new manufacturing line. The line will be installed in six months and has a budget for safety training. On the line, people must perform processes in particular ways to ensure safety. New employees need demonstrations of the correct procedures. Video, then, would be an appropriate medium because it is visual; in addition this project has a long lead time and budget, which makes video even more feasible.

In the process of organizing your information, you may find that different pieces of information are best presented in different media. For example, if you are developing documentation for a business telephone system with voice mail you could create a manual for basic use, a wallet-sized reference card for people calling into the voice mail system, and an online tutorial to teach the service representatives how to repair the telephone system.

Or, suppose you are preparing a proposal for a defense contract. You might print a written response to the proposal, addressing all of the requirements in the request. You might also present some of your proposal "live" by developing a video that highlights your technologies and their effectiveness.

Chapter 8 provides a more in-depth discussion of media.

Select Appropriate Organizational Patterns

After you have written the objectives and determined which information to include, you need to determine how to organize the information and then do so. The process of determining your purpose (described previously in the chapter) hints at the *organizational patterns* to use. For example, the questions readers asked about the swing set assembly suggested that they required a general-to-specific pattern of information with some general information up front followed by the specifics of assembly. Following are examples of organizational patterns.

Pattern	**Example**
Chronological	Technical report describing the steps in upgrading an emissions system at a power plant to meet new government regulations.
Most important to least important	Plan of action for addressing safety conditions, starting with the most important actions.
Most-used to least-used	A word processing manual discussing the tasks users perform most often (saving the file) before those they perform less often (printing multiple columns).
Familiar to unfamiliar	A technical report on digital analog technology that begins with a description of more common analog technology.
Question/answer	A book for computer novices whose first chapter answers basic questions such as "What is a computer program?"
Comparison	A feasibility study that compares the commercial viability of artificial intelligence in different applications.
General to specific	Beginning a complex multipart procedure with an overview.

(*continued*)

Pattern	Example
Problem/solution	A chart for appliance technicians that lists symptoms of problems in the left column and solutions in the right.
Task	A manual for a word processing program that provides separate procedures for each task users perform, such as cutting and pasting text.
Simple to complex	A phone system manual that describes simple uses like putting a call on hold before more complicated uses like call waiting.
Hierarchical	A report on house plants has a section called "Care," which is further broken down into sections such as "Diseases." "Diseases" is further broken down into "Pests" and "Organic Causes" and so on.

Documents can have several organizational patterns; even separate sections can have different patterns. For example, a simple marketing brochure for a technical product may order information in the brochure from most important to least important; order features from simple to complex; organize a paragraph by comparing the product with competing ones; start with general information about the product and end with specific technical specifications; and chunk the information so it can be read in a nonlinear way.

A document can have a complex mix of organizational patterns. For example, it might follow a familiar-to-unfamiliar organization, but the familiar material might be conceptual and organized in a general-to-specific pattern. The unfamiliar material might be a procedure and organized chronologically. By using the appropriate mix of organizational patterns, you ensure that each part—section, chapter, and document—conveys the information that readers need. See "An Example of the Design Draft Process" elsewhere in this chapter for an example of mixed organizational patterns.

Select an Organizational Technique for the Design Draft

Armed with a body of information and a variety ways of organizing it, you can now develop your design draft. The large amount of information, the number of choices to make, and the need for one document to satisfy several types of readers all have implications for how you organize material. Fortunately, techniques and tools exist that can make organizing your information easier, including the following:

- Outlines
- Mindmaps
- Question-and-answer
- Computer tools
- Storyboards

Outlines. Many people shudder at the word "outline." You remember elementary and high school teachers requesting structured sentence or topic outlines for projects and essays. The purpose of an outline, they said, was to clarify your thoughts and provide a plan to follow when writing. Some of you may have been graded on how well your final document measured up to your outline. Figure 2.2 shows an example of a topic outline for a full-text searching experiment.

I. Introduction
    A. Similar studies
    B. Program features
II. Method
    A. First session
        1. Pen and paper
        2. Discussion session
    B. Second session
        1. Hands-on
        2. Thinking-aloud protocol
III. Results
    A. Comments of participants
        1. Positive comments
        2. Negative comments
    B. Command usage
        1. Search commands
        2. Display commands
    C. Errors
        1. Syntax errors
        2. Logic errors
    D. Effectiveness
        1. Number of correct answers
        2. Speed of participant
IV. Discussion
    A. Improved text display required
    B. Improved user interface required
    C. Improved method of training required
V. Conclusions
    A. Full text suited to fact-finding
    B. Full text provides good display facilities
    C. Users need training in full text

FIGURE 2.2 Topic Outline for Technical Report

This view of the outline rests on the assumption that people know exactly what they want to say when they write the outline. Technical communicators know this simply isn't the case. You are constantly dealing with changing information. Your outline changes as the products you document change, as your knowledge of your reader grows, and as your view of the material changes.

Studies have shown that writing suffers when a writer creates a formal outline too soon in the writing process (Flower, 1981). When writers view the initial version of an outline as a near-final product rather than as a working plan, they write to the plan even when the circumstances change and the plan is no longer a good one. This isn't to say that you shouldn't write an outline before you start drafting. You should view the outline as a visual record of the thinking process as opposed to an outline of the finished product (Flower & Hayes, 1981).

Mindmaps. An alternative to the traditional sequential outline is a graphical outline sometimes referred to as a *mindmap* (Buzan, 1983). A mindmap is a drawing in which you place all the main ideas of your information and draw connections between them. Mindmapping lets you work out the hierarchical relationships of the information without worrying about the order.

You draw a mindmap as follows:

1. Place the main idea on the center of the page and draw a box around it.

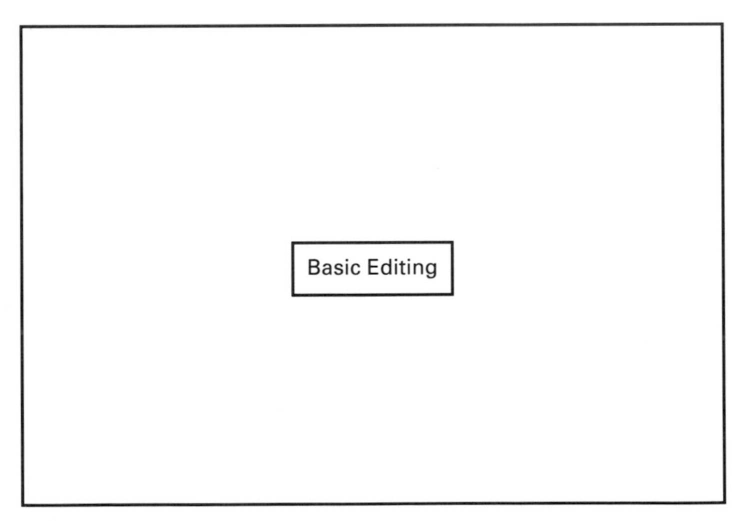

FIGURE 2.3 Main Idea

2. Place subordinate ideas around the main idea. Draw circles around these. Then, draw lines showing connections of the subordinate ideas to the main idea and to each other.

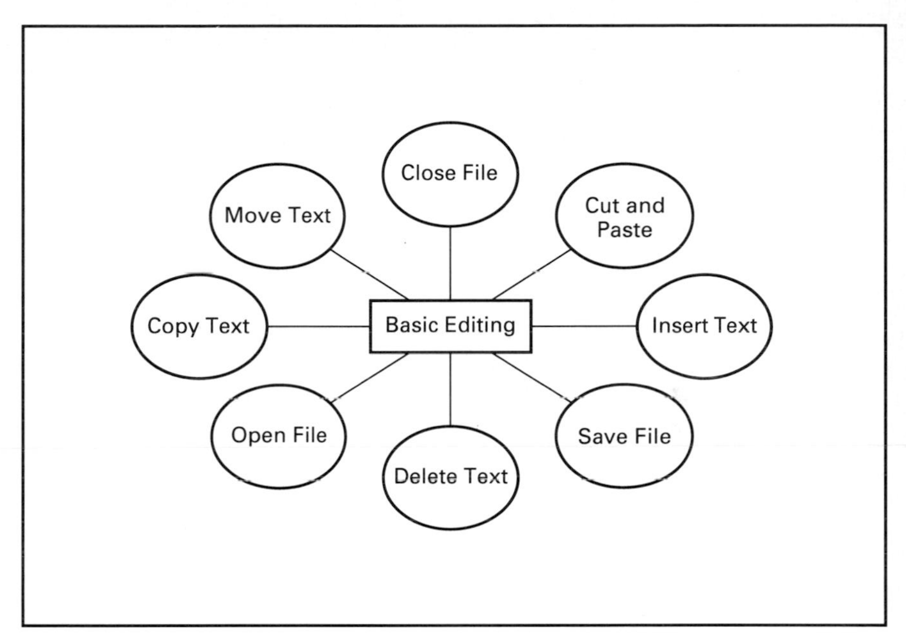

FIGURE 2.4 Subordinate Ideas

3. Continue for each additional level of detail until finished.

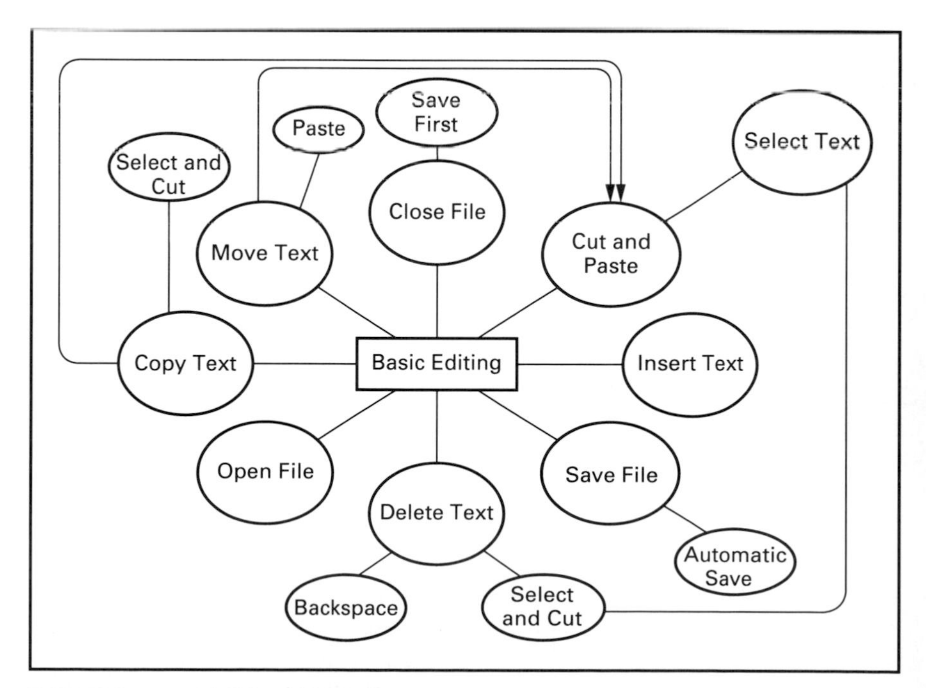

FIGURE 2.5 Additional Detail

Mindmapping offers three advantages over traditional outlining. First, the format lets you ignore the issue of order. You put down points as you think of them and worry about their order at a later stage. This allows you to concentrate on the issues and relationships between them.

Second, points that don't seem to fit can be written anywhere on the page and connected to other points later. The format of a mindmap encourages this more than the traditional outline, which encourages you through its format to write your points in order.

Third, the structure of the mindmap shows the relationships visually. The size and density of various clusters of information show you the strength of the relationships between information. This may help you identify gaps, overemphasis of information, or information that needs to be broken down further.

Mindmapping is best done on a whiteboard or blackboard, which can be easily updated as ideas change. If unavailable, flip chart paper or large sheets of bond paper can suffice. Whichever you choose, try to have all the information in front of you at once, so you visualize the whole document at a glance.

Mindmaps are especially useful when several people are working on a project. Mindmaps can help everyone keep a handle on all the information in a project. Posting the mindmap on a whiteboard, blackboard, or bulletin board keeps everyone abreast of organizational changes in the information—changes that may affect them. As a result, team members have a sense of both the document's parts and its "whole." They can then focus on their own area without losing perspective on the whole document.

Question-and-Answer. Coming up with questions that the reader will ask is another useful technique for both gathering information and for testing organizational patterns (Fawcett, Ferdinand, & Rockley, 1989). This can be effective as a step before mindmapping or outlining. For example, the technique was used to organize this chapter. First we generated a list of questions to which we expected readers would want answers:

- Why is organization important?
- What organizational patterns are important for writing documents?
- How do you formalize the organization?

Then we used mindmapping and traditional outlining techniques to create our design draft.

Figure 2.6 shows questions that aided in the organization of an installation manual.

Computer Tools. Computer tools can also help you work out the initial organization of your material. Software programs called *outline processors* let you take a top-down approach, in which you define the structure first (as you do if you start with an outline) before writing any detailed content.

What are the hardware requirements?
What are the software requirements?
How much disk space is required?
How much memory is needed?
How long will it take?
How do I install it?
How do I get help if I need it?
How do I know the installation is successful?

FIGURE 2.6 Question-and-Answer Technique

Many standard word processors have outlining features built in. These programs let you see a top-level view of your document by hiding the details. For example, the chapter headings or section headings may be displayed without the lower-level headings and text. Some also allow you to move the entire contents of chapters or sections simply by moving their corresponding headings in a top-level view. Some hypertext authoring products also have tools for organizing material. Chunks of information are developed and then connected or linked to each other to show hierarchical and associative relationships.

Storyboards. Storyboarding is a technique adopted from the motion picture industry. The story or narrative of the film is designed on storyboards. Storyboarding is not new to user documentation (Hughes Aircraft Company adopted the technique in the 1960s). The technique became popular among technical communicators when *How to Write a Usable User Manual* was published (Weiss, 1985). Weiss referred to the storyboards as *module specs.* Each *module* is a standard-sized piece of information (usually one or two pages) that addresses a single theme. A module spec describes the module. In Weiss's model, the spec contains a headline (meaningful heading), a summary paragraph, description of exhibits (illustrations, charts, etc.) and notes about the content. The following shows an example of each element:

Headline	Moving text
Summary Paragraph	Moving text is a two-step process. First, the text to be moved is highlighted and cut from the file. Second, the text is pasted back at the new location.
Description of Exhibits	Pictures of menu with "cut" option and "paste" option highlighted
Notes	Page should also explain that the text is stored to a clipboard and how users can go into the clipboard to retrieve text.

Weiss's storyboard model can be adapted to suit your own needs. Sometimes a one- or two-page module is too limiting for lengthy or complex information. For example, a short tutorial at the beginning of a user manual may require a six-page module if chunking it further disrupts the flow of the procedure. Some information is hierarchical in structure and demands that hierarchy be built into the modules (by having subheadings within a module or overview modules, for example).

Storyboarding on paper is also useful for online documents. A module spec can be written for each screen or frame. The module spec can then be assembled into a storyboard that shows the flow of the document or network of information. Like the mindmap, the storyboard can be posted on a bulletin board or divider to serve as a review draft or simply to keep everyone informed of a document's progress.

Formalize the Organization

The organization of your information is formalized in the design draft. The design draft is an outline of your information. It may look like a traditional outline or a mindmap; it may consist of a general description of the structure of the information or may be filled in with details such as sample paragraphs and questions. It may be a storyboard of each page of the document or each screen in a tutorial or online document. Several design drafts are shown below. Figure 2.7 shows a "bare bones" design draft, a preliminary table of contents for a wine-making brochure.

THE BEGINNER'S GUIDE TO WINE-MAKING

History of Wine-Making
Types of Wine
What You Need to Make Wine
Steps to Perfect Wine
  Sterilizing Equipment
  Adding the Yeast
  Fermenting the Wine
  Siphoning the Wine
  Bottling the Wine
  Storing the Wine
Selecting the Right Wine to Complement a Meal

FIGURE 2.7 Example of a Preliminary Table of Contents as Design Draft

Figure 2.8 shows a design draft represented as a mindmap. This design draft is based on the one used as an organizational technique in Figure 2.5. Note that the ideas are now expressed as headings, and the order of tasks is indicated by arrows.

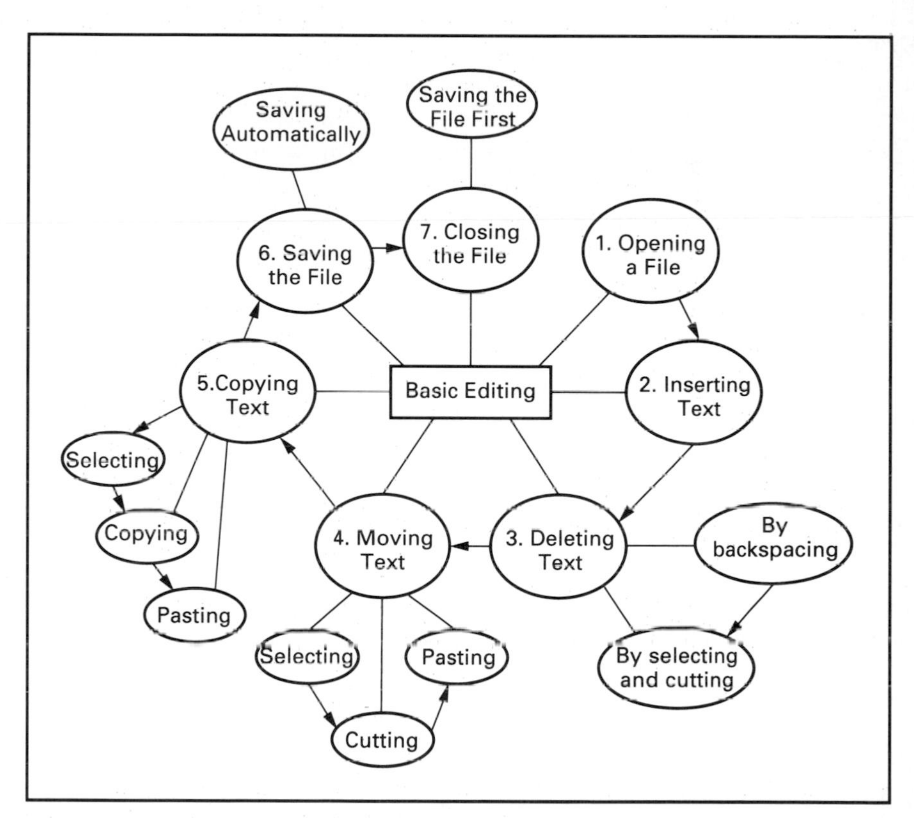

FIGURE 2.8 Mindmap Used as Design Draft

Finally, Figures 2.9 and 2.10 are detailed down to the storyboard level. Figure 2.9 shows a storyboard page from a user manual, and Figure 2.10 shows a storyboard screen from an online tutorial. The storyboard page provides information about the actual content of a specific page from the document; the tutorial storyboard screen in this case is more general. It provides information, not on the actual content of the screen, but on the positioning of its parts. This figure shows what types of information will appear on a first panel of a lesson and where each type of information will appear on the screen.

1.2. STARTING AND LEAVING SEARCHER

Page Summary:

You start Searcher by typing the name of the program and the name of the text file you want to search. To leave Searcher, type "stop", "done", or "quit". To interrupt Searcher in the middle of a search, type "Ctrl + C".

Page Content:

- list command to start and provide example
- explain the prompt
- list commands to stop and provide examples
- list command to interrupt

Questions to Reviewers:

Is there a way to stop your search session temporarily and go back to it later?

FIGURE 2.9 Storyboard of a Page from a Computer User's Guide as a Design Draft

Lesson Title Page 1 of ?

Objectives of Lesson Optional Illustration

Optional Notes

Forward Back Main Menu Help Glossary Exit

FIGURE 2.10 Storyboard of Tutorial Screen as a Design Draft

Each of the formats for the design draft–outline, mindmap, or storyboard—is suited to particular situations. A summary of the advantages and disadvantages of each format follows.

Format	Advantages	Disadvantages
Outline (Table of Contents)	Reasonably quick to develop. Quick to review. Familiar development and review tool.	Not as developed as storyboard. Easy to miss potential problems in structure. Hard to visualize the amount of information.
Mindmap	Easy to add new information to. Effective discussion tool—can be presented on whiteboard or flipchart. Reasonably quick to develop.	Hard to relate to table of contents. Unfamiliar tool for many people. Requires a large surface for presentation. Doesn't show order of information well.
Storyboard	Develops organization of information in detail. Provides substantial material for review. Identifies problems in structure early. Reflects capacity of publishing medium (e.g., page, screen). Saves time in writing further drafts. Serves as a detailed template for individual or team writing.	Time-intensive to develop. Potential for wasted effort if subject matter changes drastically. Takes longer to review because more developed. Unfamiliar tool for many people.

You should consider the specifics of your own situation when choosing a design draft format to use. Use them all to aid you in your work. For example, if your manager wants to see a working outline of your document on his or her desk tomorrow morning, provide a table of contents. You may find it helpful to use a mindmap to select the material and analyze the relationships before you formalize the organization into a table of contents. Once your initial table of contents is approved, you can go on to develop a storyboard.

Similarly, if you are documenting a subject that is unstable, you might stick to an outline or mindmap until the material is nailed down or to a storyboard for only those areas that are stable. If you try to develop the material in detail before your subject matter stabilizes, you may find yourself with a well-structured but irrelevant design.

Because it provides enough material for reviewers to analyze, a storyboard can be a good basis for review. At the same time, a storyboard does not require you to spend too much time writing polished sentences. The storyboard shows how the page, screen, or frame will look: where information will be positioned and a sample of some of the content such as headings or summaries. Storyboarding is especially important when the communication medium limits the size of the information unit, as is often the case with online documents, slide, video, or other material. The storyboard uncovers most of the problems in structure before you begin writing the first draft of the information.

Evaluate the Design Draft

The design draft represents your best attempt at planning the information before you begin writing in detail. As you become more familiar with the technical material, determine whether your first attempt is still your best attempt. You can do this by formally or informally evaluating your information early and often.

The proposed organization of your information should be evaluated throughout the writing process, from the design stage through the final draft. The following section gives you some hints on how to do this.

Evaluate the Organization Yourself. See if the organization of the information suits your readers, purpose, content, and medium. Specifically, take the list of objectives you defined for your information and

- Make sure that it includes the information necessary for readers to attain each objective.
- Identify the organizational patterns proposed for your information. Are they being used effectively? Are these patterns what your readers would expect, considering their experience level and information needs?
- Ensure that the top-level headings are parallel in structure and importance to each other. Check that subheadings are really subordinated to main headings.

Get Others to Evaluate the Organization. Get input on your proposed organization from others. The organization may be logical to you as the technical communicator but may not make sense to others. As the communicator, you may unconsciously supply information when you read your own work that is missing to readers. You may have worked with the information so long that you can no longer read it objectively.

Perform usability tests, even at the design draft stage. These are some suggestions:

- Give reviewers and readers a detailed outline, mindmap, or storyboard and ask them to check that the information flows logically. If you are documenting a system or product, ask people to check the flow of the information while using the system or product.

- Give reviewers the list of objectives for the information and ask them to check whether these have been fulfilled. If, during evaluation, your reader profile changes, revise your organization accordingly.

See Chapter 11 for a detailed discussion of planning and conducting usability tests.

AN EXAMPLE OF THE DESIGN DRAFT PROCESS

This example illustrates the process of creating a design draft for one small document (20 pages), a user guide for a product called, the "Network Expander." The example takes you through each step described in this chapter, but your analysis should be much more detailed than this one, which is provided only to give you an overall understanding of the process.

Determine the Purpose

The first step is to determine the purpose of the document by determining what you are writing about, who you are writing for, and what the objectives of the document are.

Determine What You Are Writing About. Use the following question-and-answer process.

What is the purpose of the product?	The Network Expander is a piece of equipment used on a computer network to increase the number of workstations (computers) and the total length of cable in your network system.
Who will use the product? Where will they use the product?	Computer experts called, "Site Administrators," hired by school boards will install Network Expanders in elementary and secondary schools. Occasionally, in remote locations or where resources are low, a teacher will assume the role of Site Administrator.
Why will they use the product?	The Network Expander will be used when the network is larger than 16 workstations and the cable required to hook the workstations together is more than 200 feet.

How will they use the product?	The Network Expander will be used primarily to connect two classrooms that are more than 200 feet apart. They will sometimes use the Network Expander to connect more than 16 workstations in the same classroom.

Determine Who You Are Writing For. Do this by asking the following questions.

Who are your readers?

Site Administrators:

- Hold computer science degrees
- 25 to 35 years of age
- Maintain networks of computers
- Plan computer configurations
- Plan layout of networks

Elementary and secondary school teachers:

- Generally in mid-thirties
- Have at least an undergraduate degree and a teacher's certificate but the degree is not necessarily in a discipline related to computers
- Spend most of their time teaching in the classroom.

What do readers already know?

Site Administrators already know:

- Much about networks in general
- Little about installing a network that extends beyond a single classroom

Elementary and secondary school teachers already know:

- Little about networks
- Little about how a computer works
- Something about applications on the computer
- Nothing about installing a network that extends beyond a single classroom

How are readers going to use the information?

To design the network over the summer months in preparation for September classes (reading-to-learn)

- To install the network (reading-to-do)
- To resolve problems quickly when the network fails (reading-to-do)

- To learn how to find information because the reader won't read all the information at one sitting (reading-to-access)

What questions will readers ask?

Site Administrators will ask:

- How do I design a network?
- What is a Network Expander?
- How do I install it?
- How does it work?
- When do I need to use it?
- What does it do?
- How do I fix it?

Elementary and secondary school teachers will ask:

- What is a network?
- How do I design a network?
- What is a Network Expander?
- How do I install it?
- How does it work?
- When do I need to use it?
- What does it do?
- How do I fix it?

Hints about how the information should be organized can be surmised from the reader profile. First, the characteristics of readers (who they are, what they already know) suggest two distinct reader groups. One group is teachers who are inexperienced with network expanders and networks in general; they require a lot of conceptual information to explain what a network is and what a network expander is for. Information that is organized from general to specific, familiar to unfamiliar, and simple to complex will help them deal with new and complex concepts. The other group is Site Administrators, who will not need as much conceptual or background material. They will be most interested in the installation and troubleshooting details.

Second, how readers use information provides clues to organizing the material. Since readers will be reading to learn, reading to do, and reading to assess, readers will require conceptual, procedural, and logically organized information.

Third, the questions readers ask also provide hints on organizing material. Questions beginning "What" and "How does" are requests for conceptual or general information. Again, information organized from general to specific, familiar to unfamiliar, and simple to complex is appropriate. Questions beginning "How do I" are requests for procedural information. Information should be organized chronologically or by problem. Questions beginning "When" are requests for specific information.

Determine the Objectives of the Document. The next step is to determine the objectives for the information. The questions asked by the readers suggest six terminal objectives. The terminal objectives resulting from using this document are as follows:

- Readers should correctly define the term "network."
- Readers should correctly define the term "Network Expander."
- Readers should be able to explain how the Network Expander works to a fellow teacher or site administrator and have this person relate the information back.
- Readers should be able to plan and design a network suitable to the number of workstations and size of classroom.
- Readers should be able to install a Network Expander within 45 minutes.
- Readers should be able to identify a problem with the Network Expander and solve the problem within one hour.

The terminal objectives can be broken down into enabling objectives. For example, to define Network Expander, readers must know what a network, workstation, and cable are. Thus, the terminal objective "Correctly define the term Network Expander" can be broken down into three enabling objectives:

- Define a network
- Define workstations
- Define cable

A task analysis for the terminal objective "Install a Network Expander" identifies the following tasks:

- Determining the proper location
- Assembling the tools
- Powering down the network
- Mounting the Network Expander
- Attaching the cables
- Hooking up the network
- Powering on the network
- Testing the network

The corresponding enabling objective for "Attaching the cables" would be:

- With the assistance of the document, readers should correctly attach the Network Expander to the network cable.

Select the Communication Medium

Information on installing the Network Expander cannot be presented as an online document because the computer system must be powered down to install the Network Expander. Also, Site Administrators and teachers may want to plan their networks over the summer break when they have no access to a computer system. A video would be expensive to produce and would require teachers and Site Administrators to have VCR equipment. So the Network Expander information should be printed.

Two different types of print documents are recommended:

- Operations guide primarily for teachers and less experienced Site Administrators who need the conceptual and procedural information.
- Quick reference card for Site Administrators who require only the specifics of installation.

This example focuses exclusively on the operations guide.

Select Appropriate Organizational Patterns

Based on the reader and task analysis, the operations guide describes the following high-level tasks:

- Understanding the Network Expander
- Designing the Network
- Installing the Network Expander
- Diagnosing Problems on the Network

The guide also has a glossary because many new terms are introduced.

Formalize the Organization

Once you understand the readers' needs and the type of information required, you can design the specifics of the guide, describing the design for each section in detail. This document is divided into chapters and the following structured headings are written for each chapter.

1 Understanding the Network Expander	
What is it? What does it do? When do you need one? How does it work?	Question-and-answer

Because readers will have questions about the Network Expander that need to be answered before they can go on to learn how to use the product, the question pattern is appropriate for the first section. The information within each subsection is organized familiar to unfamiliar, simple to complex, and general to specific.

2 **Designing the Network**	
Guidelines for designing a network Possible network configurations	General to Specific

Because readers can design the network in several ways, they first need to know general criteria for setting up a network, then need more specific guidelines for different types of network configurations. This type of information suggests a general-to-specific pattern.

3 **Installing Your Network Expander**	
Getting started Mounting the Network Expander Attaching the cables Hooking up the network Powering on the network Checking that the network works properly	Chronological

Because this chapter presents procedures, its organizational pattern is chronological.

4 **Diagnosing Problems on the Network**	
Flashing red light(s) Continuously lit red light(s) Flashing yellow lights	Problem/solution
5 **Glossary**	Alphabetical

When readers encounter a problem, they notice a symptom and question how to resolve the problem associated with the symptom. This pattern suggests the problem-to-solution organizational pattern. Solutions are provided as sequences of procedures to follow.

The final chapter is a glossary, an alphabetical listing of terms used in the guide. It is provided so that readers can look up unfamiliar words.

The design draft that results from this process is the preliminary table of contents shown in Figure 2.11.

1 Understanding the Network Expander

What is it?
What does it do?
When do you need one?
How does it work?

2 Designing the Network

Guidelines for designing networks
Possible network configurations

3 Installing Your Network Expander

Getting started
Mounting the Network Expander
Attaching the cables
Hooking up the network
Powering on the network
Checking that the network works properly

4 Diagnosing Problems on the Network

Flashing red light(s)
Continuously lit red light(s)
Flashing yellow lights

5 Glossary

FIGURE 2.11 Design Draft for Operations Guide

Evaluate the Organization. The preliminary outline can now be given to reviewers who should answer the following questions:

- Does the guide contain the information that readers need to achieve each objective?
- Does the organization suit the purpose, audience, content, and medium?
- Are the top-level headings parallel to one another? Are the subheadings subordinated to the main headings?

After changes are agreed upon at the review stage, a detailed storyboard can be developed.

SUMMARY

You can design your documents so that the information is easily learned and recalled by following these guidelines:

- Relate new information to what readers already know.
- Help readers draw the right inferences from the material.
- Provide readers with the gist of the information.
- Limit the amount of new information you present at a time.

Develop the design of your document as follows:

1. Begin by clarifying the purpose of the document, which involves the following steps:
   - Determine what you're writing about.
   - Determine who you're writing for; that is, develop a reader profile by answering the following questions:
     - □ Who are your readers?
     - □ What do they know?
     - □ How are they going to use the information (reading to do, reading to learn, reading to assess, reading to learn to do)?
   - What are the questions they will ask?
   - Determine the terminal objectives for the information. Identify what readers should be able to do after reading the information. Also identify the enabling objectives. Use a task analysis to help select the information necessary to obtain the objectives.
2. Select the appropriate communication medium.
   - Print
   - Online
   - Video
3. Select an appropriate organizational pattern.
   - Chronological
   - Most important to least important
   - Most used to least used
   - Familiar to unfamiliar
   - Question/answer
   - Comparison
   - General to specific
   - Problem/solution
   - Task
   - Simple to complex
   - Hierarchical
4. Then use one or more of the following organizational techniques or tools:
   - Mindmap
   - Question-and-answer

- Traditional outline
- Computer tools
- Storyboard

5. Formalize your proposed organization in a table of contents, mindmap, or storyboard.
6. Continually evaluate your design draft as you become more familiar with the project. Check that the organization of your information suits your readers, purpose, content, and medium.

LIST OF TERMS

chunking Process of breaking down information into small components.

design draft Shows the structure of the information in an outline, mindmap, or storyboard.

enabling objectives State the prerequisite skills and background information required to achieve the terminal objectives.

mindmap A visual representation of a topic or task that shows the hierarchical and associative relationships between ideas.

module A unit or chunk of information (often a two-page spread in printed documents and a screen or frame of information in online documents).

module spec A detailed plan or design of a module that might include a heading, summary, exhibits, and notes about content.

organizational patterns Common groupings and sequences of information that are chosen according to the document's purpose and its readers' needs. The general-to-specific pattern, for example, is used when teaching the reader new information.

outline processors Computer software programs that help users create and develop outlines.

storyboard A series of module specs that show the flow of information.

terminal objectives Objectives that state the broad skills and knowledge readers acquire after reading information.

task analysis An analysis of the task the reader must perform to achieve the objectives of the information.

REFERENCES

Buzan, T. (1983). *Use both sides of your brain* (pp. 5–9). New York: E.P. Dutton.

Dick, W., & Carey, L. (1990). *The systematic design of instruction.* Glenview, IL: Scott, Foresman.

Diehl, W., & Mikulecky, L. (1981). Making information fit workers' purposes. *IEEE Transactions on Professional Communication, 24,* 6.

Fawcett, H., Ferdinand, S., & Rockley, A. (1989). Organizing information. *Proceedings of the 36th International Technical Communication Conference* (pp. WE 15–18). Chicago: Society for Technical Communication.

Flower, L., Hayes, J. (1981). Plans that guide the composing process. In C. Frederickson & J. Dominic (eds.), *Writing: The nature, development, and teaching of written communication: Vol. 2. Writing: Process, development and communication* (pp. 39–58). Hillsdale, NJ: Lawrence Erlbaum Associates.

Flower, Linda. (1989). *Problem-solving strategies for writing.* San Diego: Harcourt Brace Jovanovich.

Kent, Thomas. (1987). Schema theory and technical communication. *Journal of Technical Writing and Communication,* 17(3), 243–251.

Meyer, B. (1985). Signaling the structure of the text. In D. Janassen (Ed.), *The technology of text: Vol. 2. Principles for structuring, designing, and displaying text* (pp. 64–89). Englewood Cliffs, NJ: Educational Technology Publications.

Redish, J. (1989). Reading to learn to do. *IEEE Transactions on Professional Communication, 32,* 289–293.

Weiss, E. (1985). *How to write a usable user manual.* Philadelphia: ISI Press.

SUGGESTED READING

Duin, A. (1989). Factors that influence how readers learn from text: Guidelines for structuring technical documents. *Technical Communication, 36,* 97–101.

Huston, K. & Southard, S. (1988). Organization: The essential element in producing usable software manuals. *Technical Communication, 35,* 179–187.

Meyer B. (1985). Signaling the structure of the text. In D. Jonassen (Ed.), *The technology of text: vol. 2. Principles for structuring, designing, and displaying text* (pp. 64–89). Englewood Cliffs NJ: Educational Technology Publications.

Redish, J., Battison, R., & Gold, R. (1985). Making information accessible to readers. In L. Odell & D. Goswami (Eds.), *Writing in nonacademic settings* (pp. 129–153). New York: Guilford Press.

White, F. *The writer's art: A practical rhetoric and handbook.* Belmont, CA: Wadsworth.

CHAPTER 3

PLANNING AND TRACKING A PROJECT

James Prekeges

Microsoft Corporation

JANUARY						
SUN	MON	TUES	WED	THUR	FRI	SAT
		1	2	3	4	5
6	7	8	9	10	11	12
13	14	15	16	17	18	19
20	21	22	23	24	25	26
27	28	29	30	31		

INTRODUCTION

Every technical document meets a specific need for information. Whether the document is a one-page quick reference card for filling out a claim form or a multi-volume set of manuals integrated with online, multimedia help and tutorials for a software application, the information must be provided at a certain level of quality, at a certain time, and within a certain budget.

These three factors—quality, speed, and cost—are inevitably at odds with one another. By determining the priorities of these three things as you begin a project, however, you can more easily make the right decisions for optimizing quality while minimizing time and cost. You determine these priorities, and how they affect the quality, speed, and cost of producing information, during the planning phase and track them during the development phase.

This chapter describes the process of planning and tracking a project, which is the same for all projects, regardless of the communication medium used. These general strategies need only slight adjustment to fit your specific situation.

The steps in planning and tracking a project are

1. Planning your strategy by
   a. Selecting the project team.
   b. Defining the scope of the project.
2. Developing the schedule for the project by
   a. Estimating the total time needed to complete the project.
   b. Creating a schedule of intermediate deadlines (milestones).
3. Establishing the budget for the project.
4. Planning for quality.
5. Identifying the conditions underlying the plans.
6. Documenting your management plan.
7. Tracking the progress of the project and adjusting the schedule and budget as needed.
8. Evaluating the project after it is completed to identify successes that can be repeated and problems that should be avoided.

Although this chapter does not provide estimates for schedules and budgets, it does identify the components you should consider when developing these estimates and suggests how you can estimate those components and thus estimate a schedule and budget.

STEP 1: PLAN YOUR STRATEGY

The first step in planning a project is to define its strategy or direction. Whatever project you work on, whether it is a rushed one that requires a lot of work in a short time, or a run-of-the-mill one that may be scheduled with adequate time, you need a strategy to guide you. On a personal level, what seemed obvious early in a project often becomes unclear in the heat of writing. A strategy set early in the project provides the clarification you need later on. At a broader level, a strategy also provides a vision of a project that is shared by a project team. A project team is the group of people who work on a document. When the team needs to make a decision during the course of a project, it can refer to the strategy for clarification of the project's goal and make decisions accordingly. Planning the strategy involves two steps:

1. Organizing the project team
2. Setting the initial strategy

Organizing the Project Team

Technical communicators rarely develop projects on their own; they usually work with a team of people. Some team members have technical expertise, some have communications expertise, and others bring expertise unique to the type of project you are working on.

Planning the strategy usually provides the team with its first opportunity to work together. Effective teamwork is essential to the success of any project; see Chapter 4 for a discussion of interactions among members of a team.

Also essential to the success of a project is having the right skills on the team. The skills needed on a team vary, depending on the type of project. Consider how the skills vary for teams producing product-related information, proposals, research reports, and training materials:

Type of Project	Team Member's Skill	Team Member's Responsibility
Product-related information	Marketing	Take the lead in bringing the product to market. The marketing member should have researched and addressed such issues as the need for the product, who will be purchasing it, how and where the product will be used, the knowledge and background of the users, tasks the users will perform with the product, and so on.

(continued)

Type of Project	Team Member's Skill	Team Member's Responsibility
	Technical	Develop the product. Technical skills include programming, engineering, and other scientific and technical skills. The team member writes detailed product specifications (a technical description of how the product should work and be built), prototypes, and other information about the product describing how the product should look and operate. The specifications answer such questions as, What is the product? How does it work? What will it be used to do?
	Testing	Make sure all parts of the product work, including the documentation. Testers are a good source of information throughout a project, especially for warnings about what can go wrong with the product.
	Legal	Review the information to make sure that it is fair to all parties and that the information does not make promises that cannot be kept.
	Writing and editing	Write and edit the information. Often, product-related projects require an entire library of information (as many as a million pages of work). Because no single writer can handle that volume of work, the project often includes several writers. Editors also work on the project, acting as the "first readers" who ensure the readability of the information and as copyeditors who identify inconsistencies and inaccuracies.
	Production	Prepare information for printing, for the computer screen, or for video. The production team is responsible for such tasks as typesetting information, preparing graphics, providing programming for online information (information displayed on a computer), producing videotapes, managing printing, and managing the duplication of diskettes and videotapes.
Proposals	Marketing or development	Oversee the entire proposal. In for-profit organizations, you deal with the marketing staff, who market services. In nonprofit organizations, you work with development staff, who "develop" sources of funding.

(*continued*)

	Technical, scientific, or programming	Provide technical information about the product or service described in the proposal. In for-profit organizations, the proposal usually pertains to a technical or scientific product or service. In nonprofit organizations, the proposal usually pertains to a program (hence, the term *programming*).
	Legal	Review the proposal to make sure that it is fair to all parties and that the proposal does not make promises that cannot be kept.
	Writing and editing	Write and edit the information. Often, proposals require that vast quantities of information (sometimes a thousand pages) be prepared within a short time (such as four weeks). Because no single writer can handle this volume of work, the project often includes several writers. Editors' duties on proposal projects are similar to those on product-related information projects.
	Production	Prepare information for printing, for the computer screen, or for video. Because proposals are such condensed projects and usually require a limited number of copies, publishing attractive proposals provides unique production challenges.
Research reports	Principal investigator (or project leader)	Have the "big picture" on how an individual research project fits into the other work in the field.
	Research assistants	Perform the research and serve as a source of data and conclusions.
	Editors (for scientific journals)	Determine what will be published and what is of interest to the audience of the publication.
	Reviewers or jurors	Review the information to determine whether the research is valid and whether the findings are significant enough to warrant publication.
Training materials	Instructional design	Design the instruction; that is, set the overall strategy and objectives and provide general outlines for each part of the project.
	Technical	Provide technical expertise with the subject matter but do not have training experience. These technical experts are called subject matter experts.
	Evaluation	Develop "tests" of the instructional materials to make sure that these materials meet the objectives.

(*continued*)

Type of Project	Team Member's Skill	Team Member's Responsibility
	Writing and editing	Write and edit the information according to the plan set by the instructional designer.
	Production	Prepare the workbooks, computer-based lessons, and videos to be used in the training program.

Setting the Strategy

After the team is established, you and the other team members can set the strategy. Setting the strategy involves two key decisions:

- Determining who you're communicating with and what you're trying to say
- Determining the priorities for the project

Determining Who You're Communicating with and What You're Trying to Say. This information is covered in the design draft. See Chapter 2 for a definition of a design draft and an explanation of how to prepare one.

Determining the Priorities for the Project. Everyone wants award-winning documents in a short time at little cost, but you can't have all three. Consider this old maxim, found on the walls of many project planners:

> • GOOD
> • FAST
> • CHEAP
>
> Pick any two.

Good, fast, and cheap refer to these competing priorities:

- *Schedule* (estimated time needed to complete the project)
- *Budget* (estimated funds needed to complete the project)
- *Quality* (not in the artistic sense, but in the industrial sense: meeting the objectives or goals established for the project)

You do not actually set the priorities for the project. Rather, you need to learn the priorities of the decision makers and follow their priorities when making decisions.

When determining these priorities, consider the following questions:

- How important is this project in relation to others? Determine whether work on this project should take priority over work on other projects—even if doing so might delay those other projects. If so, the schedule is the key priority because the work must be completed on time. If not, either budget or quality is the top priority.
- Which parts of this project are most important? Determine which aspects of the project, if any, can be eliminated if you do not have the time or funds to complete the project as initially planned.

Finally, note that schedule, cost, and quality cannot be equally important on each project. Although no one likes to specify preferences among these three, the decision must be made if a priority system is going to be established. Top management and key decision makers are responsible for determining these priorities.

Once you determine the priorities, let them guide your decision making. For example, if you must submit a proposal by a certain date to compete for a large contract, do not jeopardize the schedule. Try to provide the highest-quality proposal with the understanding that you do not have time for "extras" that might further enhance its quality. A good-quality proposal that gets in under the deadline has a better chance than an excellent proposal that is never seen because it missed the deadline.

Or, if your company wants to reduce the cost of printing a large manual, you might delay the project two weeks so you can take advantage of special printing rates (generally, the more time you provide a printer, the lower the printing costs).

STEP 2: DEVELOP THE SCHEDULE FOR THE PROJECT

After you have prepared the design draft and determined the priorities for the project, you begin the process of deciding exactly what the schedule, cost, and quality of the finished project will be.

At step 2, you consider the first of these three factors: schedule. Your primary goal in this step is determining:

1. The total length of the project (how much time the project will take from beginning to end).
2. Final and intermediate deadlines (when parts of the project will be completed so you can review them and determine whether you are going to complete the project on time).

The more precise your estimates of the workload, the more accurately you can determine when the project will be completed as well as which *resources* (such as labor and equipment) you need to complete the project.

Consider this example. Suppose, after assessing a project, you learn that it needs 400 hours of work. You can then determine whether the 400 hours of work will be handled by one person and be completed in 10 work weeks (each work week is 40 hours) or by 10 people in 1 work week, or some other combination of time.

The next two sections explain how to determine the total length of the project and schedule intermediate deadlines.

Determine the Total Length of the Project

The first task in scheduling a project is to determine how much time you need to develop it. In some instances, you use this information to provide a final deadline. In other instances, the organization requesting the information has determined the final deadline. You use the information about the length of the project, then, to determine how many people need to work on the project to make sure it is completed on time.

To estimate the total length of the project, do the following:

1. Refer to the detailed table of contents, mindmap, or storyboard from your design draft. See Chapter 2 for a discussion of these concepts.

   For planning purposes, the design draft needs to be detailed enough to identify sections that will be as small as one or two pages in print, a screen or two online, or a minute of finished video. The more detailed the design draft, the more accurate the estimate of the workload. An accurate estimate of workload results in more accurate estimates of the schedule and cost.

   For example, if you were developing a scientific paper on a new medical procedure, your preliminary effort might be a mindmap that identifies each major part of the procedure.
2. Estimate the number of pages, screens, or minutes of video each section requires.

   If the total is higher or lower than you expected, check the figures again. Double-check the estimate to see if you missed sections–or counted some twice. For example, if you are planning a training video on how to perform a procedure that normally takes 20 minutes to complete, and your estimate indicates only 15 minutes of video, you probably need to reexamine your estimates.
3. Identify special work included in the project, such as illustrations and photos for printed and online documents or special effects and

location shooting for videos. Anything beyond the "ordinary" usually requires special expertise to produce and, not surprisingly, extra time.

For example, if you are planning a user's guide for a desktop publishing system, you might include each screen that users see when formatting text as well as samples of final documents. The screens and a sample of a finished document are all examples of artwork.

Keep this information nearby when you prepare your estimate. In some instances, the special work can be prepared at the same time as the rest of the project but requires special scheduling to make sure it is completed at the same time as the rest of the project. In other instances, completing the special work might delay completion of the project. Someone from the production staff can help you determine which aspects of the project involve special work and how the special work might affect the schedule.

4. Estimate the amount of development time needed as follows:

Type of Document	Formula for Estimating Development Time
Print	estimated number of pages × $\frac{\text{hours needed to complete a page*}}{\text{8 hours}}$ = Days needed to complete the project
Online	estimated number of screens × $\frac{\text{hours needed to complete a screen*}}{\text{8 hours}}$ = Days needed to complete the project

*Base this number on experience with similar projects. Work rates vary among organizations, individuals, and types of projects. If you are new to the field, check with experienced coworkers to find out what the number should be.

For example, suppose you are trying to estimate the length of time needed to write a medical reference. You estimate that this printed document will have 500 pages. A coworker who wrote the last medical reference published by your company took about 6 hours to complete a page. You would use the following calculation to estimate that the project will take 375 work days to complete:

Estimated number of pages	500
Hours needed to complete a page	6

$$500 \times \frac{6}{8} = 500 \times 0.75 = 375 \text{ days}$$

You might be wondering how to estimate the number of hours needed to complete a page or screen. The only way to come up with a reasonable *hours-per-page* figure is specifically to identify all of the conditions involved. These conditions may include one or more of the following:

- The stability of the source material. Information about completed research (stable material) is less likely to require rewriting than information about products still under development (usually very unstable). The less stable the information, the more time you need for research and the inevitable rewrites due to changes.
- Your access to source material. For example, if you have access to a solid product specification and working product prototypes, you are more likely to understand what you are writing about than if you have to guess at this information. As a result, the more source material you begin with, the less time you spend searching for information and the more likely the information will be right the first time.
- Other duties. If, in addition to writing, you need to spend a lot of time on administrative matters, such as meetings, task forces, or committees, you need more time to complete the project. For example, suppose you are writing a research report on a recently completed biology study. With complete data and minimal administrative responsibilities, you might be able to complete the project at a rate of 4 hours per page. Or suppose you are asked to write a manual about a new accounting system that is still under development and you are going to be the lead writer and you have to remain on a standing committee exploring ways to reduce the cost of your products; you might need 12 hours to complete a page.

5. Translate the number of work days into business days. Business days refer to the actual number of working days needed to complete the project. If several people are working on a project, the number of business days needed to complete it will be less than the total number of days because several people can work on the project at the same time.

   - If you are the only person on the project and have no other responsibilities, no translation is needed. The number of work days is the same as the number of business days. You need to consider weekends and holidays, however, when stating a completion date for the project. For example, suppose you estimate that 15 work days are needed to complete a project. If a typical work week has 5 days and no holidays or vacations are scheduled, the project will take three work weeks (15/5 = 3).

Then allow for weekends and holidays (estimate 2.25 days for every 5 business days) to determine the total length of the project. (Make sure you schedule this time in at the beginning of a project, or you won't have it when you need it.) On longer projects, you might even schedule time for sick days so they don't affect the schedule when they are taken.

To determine the total length of the training project just described, add 2.25 days for every 5 business days. In this instance, add 2.25 × 10, or 23, days (round up) to the 50 business days. The total length of the project is 73 days, or 10.5 weeks.

- If you already know how many people are on the team but do not know how long the project should take, you need to determine how many business days are required to complete the project using people already on the team. For example, suppose you are developing a training project that you estimate needs 300 days of work and that has a project team with 6 people. Assuming the work is evenly distributed among the workers, the project can be completed in as little as 300/6, or 50, business days.
- If you already know the date when the project should be complete, you need to determine how many people must join the project to finish it on time. For example, suppose you are developing a large military proposal that needs 100 work days of effort to complete—and must be completed in 20 business days. Assuming, once again, that each person contributes an equal amount of work to the final project, you would need 5 (100/20) people.

  People do not always contribute equally to a project, however, and work efforts might not overlap. For example, the editor cannot edit until the writer has written something. Writing usually takes more time than editing. When making your estimates, then, consider the assumptions on which you base them.

6. Add a fudge factor. Rules of thumb suggest that you add 20 to 50% to an estimated schedule to determine how long the project might really take. The fudge factor takes into consideration errors in assumptions, such as the assumption that all people on a project contribute equally to its completion. See "The Hidden Danger of Adding People," for an additional fudge factor to use when adding people to a project.

   Consider, again, the training project that takes 50 business days to complete with 6 people.

   - If you add in a 20% fudge factor, the estimated time needed to complete the project is 60 days: 50 days + (20% × 50).
   - If you add in a 50% fudge factor, the estimate is 75 days: 50 + (50% × 50).

THE HIDDEN DANGER OF ADDING PEOPLE

How many people should be added to a project so you can complete it on time? Too few people on a project can't solve the problems; too many create more problems than they solve.

Additional writers will not reduce writing time proportionally. Ten writers will not complete a project in one-tenth the time that it takes a single writer to complete the same project. Every writer has a certain amount of start-up time, which is spent learning about the writing tools (such as perhaps having to learn to use a new word processor), learning about the topic (using prototypes, reading a specification, or other initial research), and generally settling into the project.

Another time-consuming task is coordinating the efforts of project team members. As soon as one additional writer joins the project, time must be spent coordinating that person's work. One of the writers—the lead writer—must coordinate everyone's work, keep track of cross-references the different writers are making to each others' work, keep the other writers aware of the other parts of the project that they can cross-reference, maintain style conventions, track everyone's progress, hold meetings, collect and collate everyone's work for reviews, return the appropriate sections to the correct writers, and estimate and schedule—just to name some of the tasks.

When adding writers, then, a general rule is to subtract 15% from both the writer's time and the lead writer's time for each writer added to the project. For example, with three writers, two are spending 85% (each) of their time on writing duties, and the lead writer 70%. With five writers, four are spending 85% (each) of their time on writing duties, the lead writer 40%.

Schedule Intermediate Deadlines

Once you have estimated the number of business days the project is going to take, you can develop a schedule of intermediate deadlines. These intermediate deadlines help you pace work on the project and provide checkpoints when the project team can assess the project and determine whether it will be completed on time, within budget, and at the desired level of quality. (Sometimes, the term *milestone* is used for intermediate deadline; the two terms are identical.) The sequence of key intermediate deadlines is called a *critical path*. The critical path includes all activities that must be completed before the project is finished.

Most technical communication projects have three major intermediate deadlines, called drafts: a first draft, a second draft (sometimes called a *beta draft*), and a final draft. Technical communication projects also go through a series of reviews, editing, and production. These events must be scheduled, too.

To develop a schedule, do the following:

1. Determine how much time is needed to develop each draft, for reviews, for editing, for production, and for other events.

   Writing

   A college professor once told his students, "Many managers share with their engineer-writers the belief that writing begins when one pulls some sheets of paper and a pencil out of the desk." This belief, of course, is silly. It ignores "ramp up" time—time spent learning the subject, determining communication strategies, thinking up good examples, attending meetings, and performing other tasks. As a result, the first draft takes two to three times longer to write than later drafts.

   When determining the time needed for the three drafts, then, set aside the bulk of the time for the first draft, and split the rest of the time evenly between the second and final drafts. Some people set aside 60% of their time for the first draft, 20% for the second draft, and 20% for the final draft. Others prefer a 50-25-25 split.

   Technical reviews

   Schedule time for technical reviews of the first two drafts. During these reviews, the technical members of the team read the draft and make sure it is technically accurate. (Because the third draft is a final draft, it does not receive a technical review.) Make sure you allow enough time for this important activity. Generally, leave a minimum of a week for all projects, but add another day for each 100 pages as the length increases beyond 500 pages. When scheduling time for reviews, also figure in time to copy, distribute, and collect the review copies. (Despite instant printing and fax, copying and distributing a large document still takes time.) Depending on the size of the project, this may add another week for each review.

   Testing

   Testing is a special type of review activity. Sometimes, it happens as part of a regular review; sometimes it is separately scheduled. In some organizations, every draft is tested. In others, only one or no draft is tested. Each testing organization has its own preference for scheduling and conducting tests. Your management can tell you the preference when you begin work on the project. See Chapter 11 for a description of testing documents, including the steps needed to plan a test.

Copyediting

Editors are the first readers of a document. They try to identify the elements that benefit and hinder the readers in their efforts. Therefore, editors should review each draft of the project, identifying problems before they become serious. This type of editing is called *substantive editing*. Substantive editing should coincide with technical reviews.

Editing must also occur after the final draft. At this time, the editor's concerns move away from broader issues of readability to more detailed issues of spelling, grammar and consistency. This type of editing is called *copyediting*. Chapter 10 discusses copyediting.

Because it occurs after the technical reviews are complete, copyediting must be separately scheduled.

Use the same formula for scheduling copyediting time that you use to determine review time.

Production

Production is the process of turning a draft document into a master copy of the final document—the copy that will be duplicated.

The time needed for production can vary from one week to four months after the copyediting is complete. The actual time depends on the communication medium used, the quality of the finished product, and the size of the finished product. For example, a small black and white pamphlet on copier paper may take only a couple of days to produce, while a four-color, profusely-illustrated book printed on a parchment-like paper takes many months.

Similarly, a short online document that will be transmitted electronically may take only a few weeks to produce, while an elaborate computer-based training course that uses artificial intelligence to analyze student answers and that will be packaged in a unique binder might take several weeks to produce. Not only does the course need to be produced and the program verified to make sure it runs without technical problems, but the binder must also be produced.

For video, the final production may take from three days to several weeks, depending on the complexity of production. For example, producing a copy of a class that has simply been filmed by a single, stationary camera may take only a day or two, while a heavily animated, high-quality training video may take four to six weeks—or longer—to produce.

Talk to your production experts to help you estimate the length of the production cycle.

Duplication — *Duplication* refers to copying and distributing the final document. For print documents, duplication is printing. For online documents, it is either an electronic transmission of information or duplication on diskettes or videotapes.

For printed documents, the time needed for duplication depends on the quality and size of the information to be printed. For online documents, the time needed for duplication depends on the means of transmission (electronic or diskette) and the amount of related packaging (such as binders and diskette cases) needed.

2. Place these events on a calendar. When doing so, begin at the ending date and *backschedule* (schedule backwards).

For example, suppose it is January 14 and you are scheduling development of a technical journal that must be completed by July 31. You have determined that 50% of the writing time will be spent on the first draft and 25% each on the second and third drafts. Your production department tells you that it will need 8 weeks for production and 4 weeks for printing. You need a minimum of 2 weeks for reviews, including copying and distribution time.

Your schedule, then, might be as follows:

July 31	Complete
July 2	Printing (notice that an extra day was added; the printing time includes the July 4 holiday)
May 3	Production (notice that an extra work day was added; the production time includes the Memorial Day holiday)
April 25	Copyediting
April 11	Final draft
March 27	Second review
March 13	Second draft
February 27	First review
February 13	First draft (notice that you have four weeks to produce the first draft; two weeks to produce the second and third ones)

A graphical representation of a schedule, such as the Gantt chart in Figure 3.1, is commonly used. Note that this scheduling process assumes that people work on this project exclusively during the weeks they are assigned to it and that conditions are ideal. They rarely are. Depending on the size of a project and the time to complete it, some steps may be collapsed together. For example, the first and second drafts of a proposal on a tight schedule might be combined into a single draft and testing might be eliminated.

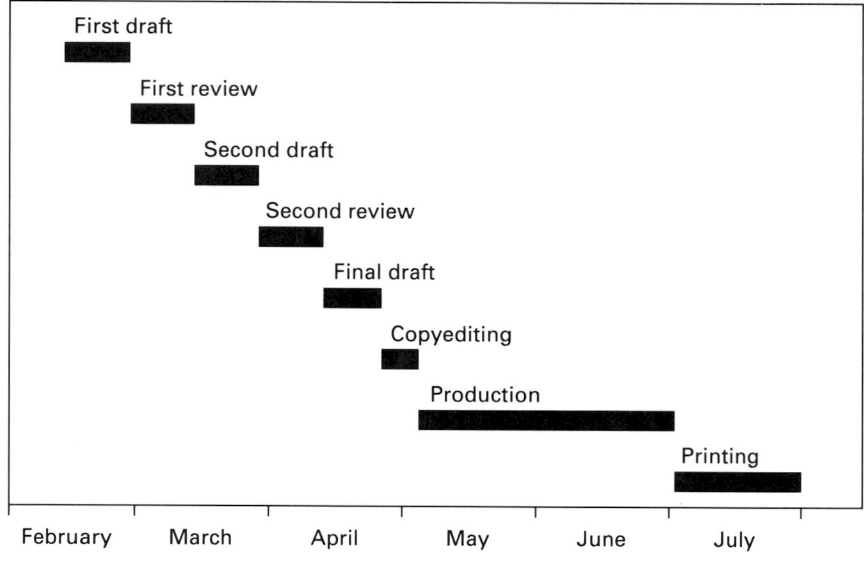

FIGURE 3.1 Gantt Chart of Project Schedule

Project management software can assist you with scheduling. With some of these programs, for example, you merely provide information about the critical path (sequence of key deadlines and the time needed between each), the number of people on the project, and the final date, and the software provides an initial estimate for you. You can then refine this estimate to meet your needs.

STEP 3: ESTABLISH THE BUDGET FOR THE PROJECT

With a strategy in mind, priorities straight, and a schedule set, you can establish the budget for the project. A budget is not the *actual* cost of the project. It is the *planned* cost of the project. You establish the budget by following these two steps:

1. Determining how much each of the components of the project (such as writing and editing) costs.
2. Determining how much you want to spend on each component. You might want to increase the expenses of some components and reduce them on others.

Later, as you receive bills for these services, compare the amount budgeted for each component with the amount actually spent.

Following are component expenses that most projects usually have:

Labor

Labor is one of the major costs of developing information. The costs for writers, editors, reviewers, and production staff are all part of the labor expense.

Because labor is such a costly item, its cost determines how many are hired for—or assigned to—a project.

To determine labor costs, use the following formula for each type of labor you plan to use on the project:

Estimated number of work days × Daily rate

(Note that these values can be expressed as hours rather than days.) You estimated the number of work days involved in an effort earlier, when you planned the schedule.

The daily rate is the daily salary or cost of a worker. Daily rates vary among workers. The rate varies for two reasons:

- Different skills command different daily rates. For example, because organizations are usually willing to pay for experience, an experienced writer usually earns more than a new writer.
- Regular employees (those who are employed on a long-term basis by an organization and receive full benefits) are usually paid a salary. Contract employees (those who are employed on a temporary basis by an organization and do not receive full benefits) are usually paid an hourly rate. That does not mean that one type of employee is more expensive to the organization than the other. Companies must set aside funds for benefits for regular employees. Companies usually pay higher wages to contract employees.

To find out the rates for various services, check with your organization's personnel department (for the salaries of regular employees) and purchasing department (for the cost of contract employees) or contact professional associations.

For example, suppose you are planning a proposal that involves 100 hours of writing, 30 hours of editing, and 60 hours of production. Suppose, too, that you plan to use contract employees to complete this project. A contract writer costs $40/hour, an editor $45/hour, and a production person, $25/hour. Your labor costs would be

Writing	100 hours × $40/hour	$4000
Editing	30 hours × $45/hour	$1350
Production	60 hours × $25/hour	$1500
Total		$6850

Equipment — Consider any equipment that must be purchased or leased to complete the project. Typical equipment includes desktop publishing systems, telephone systems, fax machines, and laser printers. You might also need furniture, such as computer tables.

Be specific about how and when each piece of equipment will be used and why it is needed. Also be clear about the tradeoff of not having a certain piece of equipment. For example, if six writers are assigned to a project, you might buy an extra laser printer for previewing review copies. Laser printers aren't cheap, but having an extra one might increase the efficiency of these writers by reducing the time they spend waiting for printouts. The cost of the increased efficiency might pay for the printer.

You can estimate equipment costs either through standard purchasing channels in your company or through a little research on your own. If you need equipment for only a short period of time (3–6 months), consider renting or leasing it.

Overhead — Overhead is the cost of operating as a business. Overhead includes expenses such as rent on the office building, lighting, heating, janitorial services, equipment already in the office, and so on.

Your organization probably has a formula for determining this cost; see your accounting department for details. You can estimate these expenses on your own by using a process similar to the one for equipment expenses.

Support services — If you need such services as production support, secretarial support, office supplies, and photocopying, account for them when preparing the budget. In some companies, support services are included in the cost of rent on office space. Otherwise, you need to determine which support services you need and investigate their cost.

Production and duplicating — Production and duplicating costs cover

- Typesetting and printing manuals and other print projects
- Preparing online information and duplicating it
- Turning a video from concept to reality and copying the videocassettes

Duplicating costs should also cover the expense of preparing review copies.

Production costs are usually the largest expense of a project and, at the same time, the most difficult to estimate because they depend on a variety of decisions.

Consider, for example, the difficulty in estimating the production costs of a sales brochure. Part of the cost depends on its size. If it is four pages, printing and production are less expensive than if the brochure is eight pages. Printing 2000 copies is less expensive than printing 3000 copies. Note, however, that the cost of printing the extra 1000 copies is not 50% more than printing the first 2000 because the most expensive part of printing is setting up the printing press, not paper or running the press. Printing one color (black ink on a white background) is less expensive than printing four-color (the effect of printing in all colors) because the paper must be run through the presses once for each of the four colors; in effect, the brochure is printed four times.

This example demonstrates the complexity in estimating print costs and how these print costs are related to other decisions about the printing process. The process is also complex for estimating the cost of online information, video, and multimedia production. For the purposes of this chapter, you only need to know that production and printing costs must be estimated. Contact a production expert to assist you with the actual estimate.

Pool all of the estimated costs into a single budget. For example, the cost of the project to develop a technical journal might be

Labor		
Writing	$ 6,000	
Editing	1,350	
Production	1,500	
Equipment	3,000	(new laser printer and fax machine)
Overhead	7,000	($1000/mo. for 7 mos.)
Support Services	3,500	(use a service—$500/mo. for 7 mos.)
Printing	10,000	
TOTAL	$32,350	

Note that these estimates are provided for example purposes only. Do not use them to construct your own estimates.

Computer spreadsheets can help you quickly estimate the cost of a project. To use a spreadsheet:

1. You provide information about each component cost and a formula for determining that cost. For example, you might determine that the formula for writing costs is hours times hourly rate.
2. You provide information for computing the formula. For example, if writing is estimated to take 40 hours and the hourly cost is $45, then the cost of writing is $40 × $45 = $1800.

Spreadsheets can also help you determine how a change in one component cost affects the final cost of a project. For example, if writing costs increase to $55 per hour, the spreadsheet can recalculate the formula for writing costs.

STEP 4: PLAN FOR QUALITY

In addition to meeting the schedule and budget, a project should meet a certain level of quality. But what is quality? Quality is an abstract concept, which each person defines in a slightly different way. To overcome this problem, you need to establish a commonly understood definition of quality for your project. This definition needs to be objective. That is, the definition needs to be stated in such a way that people can observe whether quality has been achieved. Some people define quality in terms of printing specifications (such as the type of paper used in printing and the quality of the ink) or in terms of the number of typographical errors that appear. These definitions fail to consider whether anyone can understand the information.

Another method for defining quality is borrowed from the field of instructional design: establishing observable and measurable objectives for the documentation. If objectives are written in observable and measurable terms, you can later conduct a test with readers to see whether they can meet these objectives. Chapter 2 explains how to set objectives for a document. Chapter 11 explains how to plan and conduct a test of these objectives.

Developing a design draft addresses another type of quality: *customer-driven,* or *market-driven, quality.* Customer-driven quality is concerned with meeting customers' (readers') needs. When applying customer-driven quality, you should direct your efforts to those aspects of a document that address readers' needs, not those that address your concerns or are merely cosmetic. For example, if readers like graphics but *need* pull-down menus, your first priority should be pull-down menus; you should work on graphics only if time permits. As part of the process for developing the design draft, you determine which information readers need to accomplish the goals as well as the presentation and medium best suited to achieving those goals.

STEP 5: IDENTIFY THE ASSUMPTIONS UNDERLYING YOUR PLANS

A project management effort requires resources that are often beyond your control. For example, you might need assistance from other divisions of your company and from outside video producers. Or you might need access to a restricted laboratory to try out the information you are writing.

When you develop your plans, you assume that everything will work smoothly. You assume that you will gain access to resources outside your control. You also assume that no one will be ill, that the technical information will not change, and that the staff will remain the same—no one will be replaced. Any change in these conditions could affect the schedule, the cost, the quality, or a combination of the three.

A good way to start a project, then, is stating these assumptions at the beginning so that everyone is aware of them and how changes in the conditions stated could affect the project. Identify the following when stating assumptions:

- Anything you expect from other groups, such as a prototype of the product at a specific stage, research results by a certain date, usability testing support at a certain time, specialized tools, and so on. Be specific. If you say you are expecting a "working prototype," define exactly what that means. If you don't identify what you need, you won't be able to verify that you received it.
- Anything you expect to provide to other groups, such as review copies. What are you going to deliver on the dates identified in the schedule? Be specific about your deadlines. If you are going to deliver a first draft, define exactly what a first draft is.
- The number of people working on the project and their roles in the project. If people don't know what's expected of them, they can't do it.
- Equipment you need, and when you need it. For example, if you plan to provide review copies that closely approximate the final manual, you need a laser printer well before the first draft is scheduled to go out. You might need a large computer system to store an online information project, or you might need a certain type of equipment to produce video animation.
- Anything that you have taken into account in creating the schedule. You don't want to overload people with trivial details, but you do need to let them know all of the key factors that affect your schedule. For example, you might need to say your group cannot accept other work while working on this project.

One of the best things to do when discussing these assumptions is to tie them to deadlines in the development schedule. For example, if you are writing a software manual, tie each draft of the manual to the delivery of each version of the software. Likewise, you might provide rough cuts of each section of a training video as the course is developed. In this way, your project progresses at the same rate as the technical project it supports, and if the technical project falls behind, the documentation project falls behind a similar length of time.

STEP 6: DOCUMENT THE MANAGEMENT PLAN FOR THE PROJECT

Once you have defined the assumptions, established a schedule and budget, defined quality for this project, and identified the assumptions underlying the project, record the information in a management plan for the project.

The management plan covers the following information:

- Schedule
- Budget
- Quality definition
- Assumptions

You compile this plan from the schedule, budget, quality definition, and assumptions you recorded earlier. Distribute the plan to all members of the project team.

If you are working on a small project, the management plan is usually a memo. If you are working on a larger project, the management plan is usually included in a larger document that also describes the nature and scope of the project. This larger document, sometimes called an *information plan,* includes

1. Project overview:
   a. Description of the technical information
   b. Description of the intended readers and how they might use the information
   c. Description of competition (commercial products only)
   d. Relationship to existing projects (if appropriate)
2. Design draft (see Chapter 2 for more information about this)
3. Management plan

You should review the information plan with the people who are expected to implement it before issuing "final" plans. By inviting the project team to review the plan, they are more likely to "buy into" it and perform the tasks described in it.

Note that the management plan is just that—a plan. Like all plans, this management plan might be changed after the project begins. That's fine. The management plan is a working document. You developed it to clarify the direction of the project. If the direction of the project changes, the plan should reflect this change.

STEP 7: TRACK PROGRESS

After you have made the commitment to the management plan and started the project, scheduling and budgeting tasks continue. To make sure the project meets planned schedules and budgets, take a couple of minutes each day to note what you did that day and how much time you spent on each task.

Some teams need to gather data on work for immediate needs because they charge other groups for all work time. For example, if you are a contract writer working on three projects, you need to track the amount of time spent on each project to make sure that the customer receives the correct bill.

All teams need to gather data so that they can more accurately understand how much time is needed to complete a project. Nothing fancy is required; even tracking three or four categories of activities is better than no tracking at all.

For example, a writer might track time spent on writing, research, meetings, and general administrative tasks as follows:

Activity	**Time**
Writing	40 hours
Research	10 hours
Meetings	6 hours
Administrative Tasks	5 hours

An editor might use the same categories, but track editing rather than writing. The categories you use depend on what your specific job is, what your duties are, and so on. But *do* track your time.

Tracking time not only gives you specific data for future estimates, but also shows how you are spending your time now. You can use time tracking to document interruptions that may cause the schedule to change and to decide, as the project goes on, whether you need to revise your original estimates.

Finally, make sure you track costs just as closely as you track time. As you track your progress, you may notice that you are below or above the budget. If so, re-examine your estimates and your assumptions and, if needed, revise the budget.

When investigating why actual expenses differ from planned ones, find out what happened. If appropriate, develop plans for addressing the change and re-estimate the schedule and budget for the project.

Publish your revised plan and go through the process of gaining approval for it. You are better off disappointing others early than missing a key deadline later when you do not have time to fix problems.

Some other issues to consider when tracking the progress of a project include

- Using a computerized scheduling program. If your project has a lot of people to track, is a long-running project, comes with a lot of assumptions, or undergoes frequent changes, a software program that handles scheduling can be a valuable tool. The software can quickly and accurately track the activities of many people and departments, calculate rapidly, and let you analyze the effect of changes on the final deadline.

  If your project is simple or involves just a few people, a computerized program might not be worth the effort required to enter data. In this case, you might find that using the program is taking more time than it is saving. Never depend entirely on the software. Look at the results with a critical eye because you could have entered some important data point wrong, which could throw off everything calculated with the data.

- What happens if the schedule changes? If the schedule changes, take time to rework all the deadlines and notify everyone involved in the project. Remember that schedule changes have a ripple effect. Check with the editor(s) and production people. They have schedules, too, and may not be able to accommodate your schedule change. For example, suppose a printer has scheduled your job to be printed at a certain time but you can't provide camera-ready copy at that time because the schedule was moved back one week. The printer might not be able to accommodate your new schedule because the printing presses have been scheduled for other jobs.

  Consider the effect of a supposedly small change: adding a new parameter to a software command. All cross-references in the manual must now be changed because the additional information caused all the other information to "roll over" onto other pages. All artwork showing the help screen for the original command must be redone. All descriptions or procedures that refer to that command or the menu it is in must be fixed. This simple change is not so simple. (Spreadsheets can help you determine the effect of some of these late changes.)

STEP 8: EVALUATE THE PROJECT

At the end of every project, take some time to analyze what happened, focusing on how well the project was planned and managed. The entire project team should be involved in this meeting, which is often called a *post mortem.* The discussion should be candid, and participants should be open to critical comments so that the team can benefit from lessons learned on this project for future projects.

Some of the issues to address are outlined below:

Aspect of Project	Issues to Consider When Evaluating
Schedule	Did the project take as long as you expected? More? Less? Were your hours/page figures correct? Did special work absorb more time per page than writing? If several writers worked on a project, how much time was spent supervising and coordinating their efforts? How well did you meet the schedule? Which unplanned events affected the schedule? Could you have planned for them?

(continued)

Budget	How close was the budget in estimating actual expenses? Which expenses did you accurately estimate? Which ones did you miss? Why did this happen?
Quality	Were readers able to achieve the objectives? Did achieving the objectives satisfy readers or were certain needs overlooked when the objectives were originally developed? How well did the finished project satisfy readers?
Assumptions	Were your assumptions accurate? Which assumptions that should have been stated did you overlook?
Project tracking	Did you gather enough data when tracking time and costs to better estimate the next job? If not, what else should you have tracked?
Teamwork	How well did the team work together? How could you work more effectively in the future?
Overall project	What was the best aspect of the project? If you could redo one part of the project again, what would it be?

Using the group's discussion as source material, develop recommendations for improving future projects. You might document these recommendations in a memo.

SUMMARY

This chapter presented the following eight-step process for planning, tracking, and analyzing a project. The process is a general set of procedures you can follow regardless of the type of document you are developing or the type of organization in which you work.

1. Planning the strategy:
   a. Organizing the project team. Members of the project team vary, depending on the type of project you are working on. Most teams include writers and editors, production staff, marketing staff, and technical experts.
   b. Defining the scope of the project by
      - Determining who you're communicating with and what you're trying to say (You should have determined these when developing the design draft.)
      - Determining the priorities for the project: the relative importance of schedule, price, and quality.
2. Developing the schedule for the project by
   a. Projecting the total time needed to complete the project by
      - Estimating the size of the project in pages, screens, or finished minutes of video

- Using this calculation to determine the total length of the project

Type of Document	Formula for Estimating Development Time
Print	estimated number of pages × $\frac{\text{hours needed to complete a page*}}{\text{8 hours}}$ = Days needed to complete the project
Online	estimated number of screens × $\frac{\text{hours needed to complete a screen*}}{\text{8 hours}}$ = Days needed to complete the project

*Base this number on experience with similar projects. Work rates vary among organizations, individuals, and types of projects.

b. Creating a schedule of intermediate deadlines (milestones). These deadlines include
   - First draft
   - First review (which might include an editorial review)
   - Second draft
   - Second review (which might include an editorial review and a test of the document with users)
   - Final draft
   - Copyediting
   - Production
   - Duplication

3. Establishing the budget for the project:
   a. Determining how much the components of the project cost. Components include
      - Labor (such as writing and editing)
      - Equipment
      - Overhead
      - Support services
      - Production and duplication

   b. Determining how much you want to spend on each component by pulling all of the estimated costs together into a single budget.
4. Planning for quality. If you consider the objectives for the document stated in the design draft as the quality objectives, then tests conducted later in the process can determine the quality of the document.
5. Identifying the assumptions underlying the plans, such as the availability of staff and information.

6. Documenting your management plan. The plan includes the proposed schedule and budget and covers quality and assumptions. The plan should be reviewed by the project team. The plan might also be included in an information plan, which also includes a design draft.
7. Tracking the progress of the project and adjusting the schedule and budget as needed.
8. Evaluating the project after it is completed to identify successes that can be repeated and problems that should be avoided.

LIST OF TERMS

backschedule Begin scheduling the last deadlines first to make sure that adequate time is available at the end of the project.

budget Estimate of the funds needed to complete the project

copyediting Editorial review that focuses on sentence-level issues, such as spelling, grammar, and punctuation.

critical path The sequence of key intermediate deadlines. It includes all activities that must be completed before the project is finished.

customer-driven quality A type of quality that is concerned with meeting customers' (readers') needs. Also called market-driven quality. *See also* quality.

duplication Copying and distributing the final document. For print documents, duplication is printing; for online documents, it is either an electronic transmission of information or duplication on diskettes or videotapes.

hours per page (screen) An estimated number of hours needed to complete a page (screen). Used to calculate the total length of a project.

milestone Intermediate deadline in the schedule for completing a project.

post mortem Meeting held after a project is completed to identify its strengths and weaknesses and recommend improvements for planning, managing, and tracking future projects.

production Process of turning a draft document into a master copy of the final document—the copy that will be duplicated.

quality Meeting the requirements—or stated goals for a given project. *See also* customer-driven quality.

resources Labor, equipment, and other material needs of a project.

schedule An estimate of the total time needed to complete a project, with intermediate deadlines.

substantive editing Editing early drafts of a document to identify and resolve problems that might impede the reading process.

SUGGESTED READING

Barakat, R. (1991). Storyboarding can help your proposal. In *Writing & speaking in the technology professions: A practical guide* (pp. 163–168). Piscataway, NJ: IEEE Press.

Belasco, J. (1990). *Teaching the elephant to dance: Empowering the organization.* New York: Crown.

Bell, P., & Evans, C. (1989). *Mastering documentation.* New York: John Wiley & Sons.

Benson, T. (1991). Challenging global myths: International quality study. *Industry Week, 240*(19), 12–25.

Bormann, E. (1990.) *Small group communication: Theory and practice* (3rd ed.). New York. HarperCollins.

Caernarven-Smith, P. (1990). Annotated bibliography on costs, productivity, quality, and profitability in technical publishing: 1956-1988. *Technical Communication, 37*(2), 116–21.

Dick, W., & Carey, L. (1990). *The systematic design of instruction* (3rd ed.). New York: HarperCollins.

Domenico, A. (1989). A systematic approach for developing a multi-volume documentation system. *Proceedings of the 36th International Technical Communication Conference.* (pp. MG-67–70). Chicago: Society for Technical Communication.

Edwards, A. (1989). A quality system for technical documentation. *Proceedings of the 36th International Technical Communication Conference.* (pp. MG-43–46). Chicago: Society for Technical Communication.

Frame, J. (1987). *Managing projects in organizations: How to make the best use of time, techniques, and people.* San Francisco: Jossey-Bass.

Hackos, J. (1989). Documentation management: Why should we manage? *Proceedings of the 36th International Technical Communication Conference.* (pp. MG-15–16). Chicago: Society for Technical Communication.

Horton, W. (1991). *Designing & writing online documentation: Help files to hypertext.* New York: John Wiley & Sons.

House, R. (1988). *The human side of project management.* Reading, MA: Addison-Wesley.

Killingsworth, M. J. & Jones, B. (1989). Division of labor or integrated teams: A crux in the management of technical communication. *Technical communication, 36*(3), 210–221.

Miller, P. (1989). Computerized management of time, tasks, and priorities. *Proceedings of the 36th International Technical Communication Conference,* (pp. MG-64–66). Chicago: Society for Technical Communication.

Pfaffenberger, B. (1992). *How to computerize your small business.* Carmel, IN: Que Corp.

Price, J. (1984). *How to write a computer manual: A handbook of software documentation.* Menlo Park, CA: Benjamin/Cummings.

Riney, L. (1989). *Technical writing for industry: An operations manual for the technical writer.* Englewood Cliffs, NJ: Prentice Hall.

Samuels, M. (1989). *The technical writing process.* New York: Oxford University Press.

CHAPTER 4

WORKING WITH PEOPLE

Carol M. Barnum

Southern College of Technology

No matter how proficient you become at audience analysis, project planning, writing and editing, and all other aspects of technical communication, if you don't become effective in working with people, you won't succeed as a technical communicator. Why? Because technical communicators must work with other people as a vital part of their jobs—as boundary spanners, as collaborative writers, as project planners, and as small group communicators. This chapter explores the working relationships between communicators and their project teams. Specifically, this chapter presents

1. The ways in which technical communicators work in groups,
2. The difficulties that working in groups pose for typical technical communicators,
3. The role of personality type in group interaction, and
4. The skills needed to succeed—and be happy—on the job.

TECHNICAL COMMUNICATORS AS COLLABORATIVE WRITERS

In the Introduction, Saul Carliner and I defined technical communicators as "translators" of information from those who have the knowledge to those who need the knowledge. As translators, technical communicators are *boundary spanners* (Harrison & Debs, 1988, p. 12), a role that requires that they cross organizational lines and often communicate outside the organization to both get and give information. During the writing process itself, the technical communicator interacts with "product, design, and research groups, including scientists, engineers, and programmers; corporate units such as legal, accounting, and marketing departments; and individuals in various other roles, including consumers, other writers and editors, graphic artists and representatives (Harrison & Debs, p. 12).

Certainly, it comes as no surprise to seasoned technical communicators that we work as part of project teams and engage in *collaborative writing*. However, it often comes as a surprise to the novice technical communicator because, until recently, preparation for this collaborative aspect of a professional's writing life has not been addressed in college.[1] How much do technical communicators collaborate? In a survey

1. Two recent sources dedicated to the subject indicate the academic interest. See the 1990 Special Issue—"Collaborative Writing in Business Communication" in the *Bulletin of the Association for Business Communication* (Shirk, 1990) and the 1989 *Collaborative Technical Writing: Theory and Practice* in the ATTW Anthology Series (Louth & Scott, 1989).

of 200 college-educated professionals, Faigley and Miller (1982) learned that 73.5% write collaboratively. Similarly, in a survey of writers in seven professional organizations, including the Society for Technical Communication, Ede and Lunsford (1985, 1990; Lunsford & Ede, 1986) report that 87% of the 530 respondents sometimes write collaboratively and that they collaborate on every type of document. While the extent of collaborative writing on the job is questioned by Couture and Rymer (1989) and Smit (1989), Ede and Lunsford conclude that a lot of people don't understand the term collaborative writing, and in fact many more are doing it than calling it that. They present the following seven scenarios to encompass a broad definition of collaborative writing (Ede & Lunsford, 1990, pp. 63-64):

1. Team or group plans and outlines. Each member drafts a part. Team or group compiles the parts and revises the whole.
2. Team or group plans and outlines. One member writes the entire draft. Team or group revises.
3. One member plans and writes the draft. Group or team revises.
4. One person plans and writes draft. This draft is submitted to one or more persons who revise the draft without consulting the writers of the first draft.
5. Team or group plans and writes draft. This draft is submitted to one or more persons who revise the draft without consulting the writers of the first draft.
6. One member assigns writing tasks. Each member carries out individual tasks. One member compiles the parts and revises the whole.
7. One person dictates. Another person transcribes and revises.

Debs (1991) takes this definition to its broadest extent, defining collaborative writing as "any 'co-labor' that occurs in the production of a document" (p. 479).

Even more telling than the extent of collaborative writing in the Ede and Lunsford surveys are the responses to the open-ended questions, which present a profile of effective collaborative writers as "flexible; respectful of others; attentive and analytical listeners; able to speak and write clearly and articulately; dependable and able to meet deadlines; able to designate and share responsibility, to lead and to follow; open to criticism but confident in their own abilities; ready to engage in creative conflict" (1990, p. 66). As this profile indicates, writing is only one part, and a small part, of the overall picture.

PROJECT TEAMS AND THEIR CHARACTERISTICS

Several models exist for structuring project teams, which are a type of collaborative group. Stratton (1989) describes three such models:

- The *horizontal model,* most commonly found in industry, follows the divide-and-conquer approach. Each team member takes a segment of the project and writes and edits only that part.
- The *sequential model* is characterized by duplication of effort in which one person writes a draft, then passes it to another who revises and rewrites it, passing it to a third person who again rewrites it, with perhaps still a fourth person reading and returning it to the original writer for revisions.
- The *vertical or "stratification" model* operates on the principle of division of labor. Tasks are divided according to job function, such as project manager, researcher, writer, editor, graphics person, letting each person work within an area of specialization.

Stratton recommends replacing the first two models with the third, which is common in technical communication groups.

Lunsford and Ede (1990) present another approach for structuring teams. They propose the *dialogic model,* in which the roles are flexible and may shift as the project progresses. In this approach, the process of interaction is as important as the product that the team produces. Lunsford and Ede characterize this model as "feminine" in its fluid nature as opposed to the hierarchical or more rigidly structured approaches of the traditional models. They note, however, that the more traditional models are perceived—even by the predominantly female Society for Technical Communication respondents—as "essentially unproblematic and 'the way things are'" (p. 237).

Lay (1989) also compares the behavior of men and women in groups, arguing that men and women must learn how to become androgynous; that is, each should take on the strengths of the other to improve the process of communication. Women tend to emphasize the process and the "interpersonal web" needed, men tend to emphasize problem solving and the product. Valuing both will provide greater satisfaction in groups as well as the development of successful products.

Computers and Project Teams

Experienced collaborators, according to Allen, Atkinson, Morgan, Moore, and Snow (1987), describe many models for working together on projects, but stress that the influence of computers has had a marked impact on the way in which writers can and do collaborate. Not only have computers made it easier to write, but they've also made it easier to collaborate. Whether the method employed is that of copying a disk to allow for easy editing by another or networking computers to share information at different terminals, collaborative writing has been improved almost from the beginning of computer technology. As Halpern (1985) writes, frequent requests for training in word processing at Boeing were for the very reason that the computer allows for collaboration with other writers and with the graphics people. But that's old news.

The news now is *shareware* and *groupware,* the software that is being developed specifically to enhance collaborative writing. This software is of two types:

- *Asynchronous,* allowing one person at a time to have access to a document, and
- *Synchronous,* allowing multiple users to have access to the same document at the same time.

Naturally, the excitement is over the synchronous systems, which permit interactive editing. One such synchronous system under development is the Capture Lab (Elwart–Keys & Horton, 1990), a project at the University of Michigan. This system allows a group in a conference room to discuss, write, and revise a document interactively from their chairs, as the work is displayed on a large monitor at the head of the conference table.

THE CHALLENGES OF COLLABORATIVE WRITING

"Without significant practice in working with others, people are not professionally prepared to be technical communicators" (p. 77), write Raymond and Yee (1990). A host of others could be quoted making similar statements concerning the difficulty arising from this lack of preparation. Farkas (1991, p. 6) lists six reasons why collaborative writing is difficult:

1. Highly integrated documents are very complex artifacts.
2. The process of preparing a document becomes more complex when it is performed collaboratively.

3. The writing process generates strong emotional commitments.
4. Documents are reworkable and are subject to infinite revision.
5. Collaborative writers lack fully adequate terms and concepts with which to create a clear and precise common image of the document they wish to produce.
6. It is difficult to predict or measure success.

Two studies (Doheny-Farina, 1986; Cross, 1990) point out how much can go wrong when groups don't function as they should. Doheny-Farina studied an "emerging organization" and the conflict and leadership struggle over the writing of one very important document: a business plan to be used to attract investment capital. In this study, a strong president was pitted against his group of vice presidents over the direction the company should take and the message the document should express to convey that direction. The struggle came down to one single paragraph and the different viewpoints brought to the surface in the statements made in that paragraph. Although the team did reach a compromise in the wording of the document and the company subsequently secured the funds it needed to survive, the study points up the many real social and organizational difficulties that are often an integral part of the writing process.

Cross focused on the social implications of a group-written document using the sequential model. The technical writer tried to write the letter for the vice presidents' approval and the president's signature. The technical writer, however, was not included in the meetings in which the document was discussed or in any subsequent meetings in which the drafts were analyzed. After numerous failed attempts over 77 days by the technical writer and several self-styled editors to draft the executive letter for the annual report, the senior vice president finally wrote the letter the president wanted, and it was quickly approved. Cross's study points up the problems of an unsuccessful collaboration with the result being a poor product that does not function as a clear communication tool.

Collaborative writing projects need not be as difficult as those in the two studies described. Nonetheless, technical communicators need to understand that writing is more than putting words on paper, whether done by one or many. The social implications—the hidden agenda, if you will—of all the forces at work must be recognized if technical communicators are to be successful with the process as well as the product.

Proposal writing is an excellent case in point because it often brings together a large number of people who have not previously worked together and whose goal is to produce a successful document within a specific time (usually too short). "Collaboration in a Pressure Cooker" (Bacon, 1990) is an apt title for an article that points up the problems communicators face in this very common and exceedingly important type of collaborative writing.[2] Bacon notes the problems that can result when people are "compelled by circumstance and assignment" to work together as a project team on a product fundamental to a company's survival. For instance, as Bacon points out, the team leader may have no formal authority over the people on the team, and team members may have varying levels of commitment depending on the support or lack of it from their own organization and their other job responsibilities. In my own experience working on projects for a large aerospace manufacturer, I have seen these same leadership problems and varying levels of commitment from team members, some of whom wouldn't even attend regular meetings because their managers didn't make it a priority.

Keller and Knapp (1991) drew similar parallels in a presentation given to a technical communication class, in which they offered the following scenario:

> You are a technical writer in your company who has been assigned to organize and manage a writing team for a short-turnaround technical document. You have access to a work and planning area that is equipped with several personal computers (not in a network). However, many of your expected authors are located in another building and can't be committed on a 100 percent basis to the writing effort. Most of those who can be fully committed are either within six months of retirement or are recognized as "second teamers" in their department. Most of the expected authors from your company are technically competent but possess varying writing skills—from poor to good. The writing team will also need the input of several authors from an out-of-state vendor; they pledge their total support but they would prefer writing at their own facility. Most of the authors have experience with, and access to, a word processing program, but there is no one common software program known by all authors. Some still write by hand and refuse to learn anything about computers.

Thinking about or perhaps even discussing with classmates or colleagues the issues raised in this scenario, along with possible solutions, can point up a number of the challenges facing technical communicators in collaborative writing endeavors.

---

2. See also Samson (1989).

WHY PROBLEMS EXIST IN PROJECT TEAMS

To understand why problems exist in project teams, you need to understand group process. Fischer (1991) describes group process as "the intangible part of producing a tangible product. It is learning how to perceive and react maturely to messages that are wrapped and delivered in behavior" (p. ET-113). Why do others act the way they do in groups? Why do *you* behave the way you do? To become effective in groups you need to know what motivates others as well as yourself. You need to understand people. Fischer (p. ET-113) states that we have "to pay close attention to the way in which people

- Respond to each other;
- React to frustration;
- Use their personal power to influence outcomes;
- Support or irritate each other;
- Commit themselves; and
- Are motivated or are disinterested."

Hirokawa and Gouran (1989) add that groups encounter interrelational problems, including "pressure for uniformity and deviance, authority relations, status differences, and a variety of conflicts attributable to the incompatibility of group and individual goals as well as members' roles" (p. 83).

By studying your role in groups as well as the roles of others, by learning the positive and negative aspects of conflict, by learning how to read nonverbal communication, and by understanding leadership styles, you can understand what contributes to an effective communication team.

Personality Type and Technical Communicators

To begin the process of understanding people, you have to start with yourself. Who are you? The *Myers-Briggs Type Indicator* (MBTI) is a widely used test of personality type, which categorizes individuals within 16 personality groupings, based on the work of Carl Jung. The test indicates which combination of four sets a person is by selecting one of the two letters in each of the four categories as follows:

1. Extrovert (E) or Introvert (I)
2. Sensing (S) or Intuitive (N)

3. Thinking (T) or Feeling (F)
4. Judging (J) or Perceiving (P)

Table 4.1 shows the effects of the combinations of all four preferences.

Hackos (1988), reporting on the results of testing 116 technical communicators from a mailing list of 300 plus 20 from her own company, found that the largest percentage (22%) were INTJ—Introverted, Intuitive, Thinking, and Judging. As Table 4.2 shows, this group includes managers, visual artists, writers, and editors. When the writers are isolated, the INTJ percentage rises to 45%.

Hackos indicates interest in the test because of its potential for helping us understand the impact of personality patterns on the job of the technical communicator. The technical writers in her study group were introverted types, where the majority of American society (75%) is extroverted.

Sides, who has also used the MBTI, describes the importance of understanding personality types for technical communicators because of the collaborative aspects of the job (1988, 1989, 1991).[3] Introverted types, for instance, prefer working alone in reflective, thoughtful, "withdrawn" activities. Each personality type tends to have work habits like others of the same type but different from those of other types. Moreover, each type may approach problem solving differently and even value different things more or less highly than people of other types.

Sides documents well what can happen when different types collide (1991). Working with a hardware technical writing group that was not making progress towards the completion of the task because individuals were at cross purposes, Sides administered the MBTI. His interpretation of the results helped him divide the group into a stratification model in which each person could perform from his or her strength.

While the MBTI is a popular test that can help provide some information about different personality types, we should not rely solely on the results of this or any other test in making decisions about group interactions.[4] Nor would we want to use the results to stereotype people. Learning group skills and getting good at reading others, especially those with personalities and work habits different from ours, requires keen powers of observation. Sharpening these skills is critical to becoming effective in groups.

---

3. See also Collins (1989) for an introductory discussion of type.
4. For a discussion of another aspect of communication behavior and its impact on technical communication, see Hickman (1991).

TABLE 4.1 Effects of the Combinations of All Four Preferences

		Sensing Types		Intuitives	
		With Thinking	With Feeling	With Feeling	With Thinking
Introverts	Judging	**ISTJ** Serious, quiet, earn success by concentration and thoroughness. Practical, orderly, matter-of-fact, logical, realistic, and dependable. See to it that everything is well organized. Take responsibility. Make up their own minds as to what should be accomplished and work toward it steadily, regardless of protests or distractions. Live their outer life more with thinking, inner more with sensing.	**ISFJ** Quiet, friendly, responsible, and conscientious. Work devotedly to meet their obligations and serve their friends and school. Thorough, painstaking, accurate. May need time to master technical subjects, as their interests are not often technical. Patient with detail and routine. Loyal, considerate, concerned with how other people feel. Live their outer life more with feeling, inner more with sensing.	**INFJ** Succeed by perseverence, originality, and desire to do whatever is needed or wanted. Put their best efforts into their work. Quietly forceful, conscientious, concerned for others. Respected for their firm principles. Likely to be honored and followed for their clear convictions as to how best to serve the common good. Live their outer life more with feeling, inner more with intuition.	**INTJ** Have original minds and great drive which they use only for their own purposes. In fields that appeal to them they have a fine power to organize a job and carry it through with or without help. Skeptical, critical, independent, determined, often stubborn. Must learn to yield less important points in order to win the most important. Live their outer life more with thinking, inner more with intuition.
	Perceptive	**ISTP** Cool onlookers, quiet, reserved, observing and analyzing life with detached curiosity and unexpected flashes of original humor. Usually interested in impersonal principles, cause and effects, or how and why mechanical things work. Exert themselves no more than they think necessary, because any waste of energy would be inefficient. Live their outer life more with sensing, inner more with thinking.	**ISFP** Retiring, quietly friendly, sensitive, modest about their abilities. Shun disagreements, do not force their opinions or values on others. Usually do not care to lead but are often loyal followers. May be rather relaxed about assignments or getting things done, because they enjoy the present moment and do not want to spoil it by undue haste or exertion. Live their outer life more with sensing, inner more with feeling.	**INFP** Full of enthusiasms and loyalities, but seldom talk of these until they know you well. Care about learning, ideas, language, and independent projects of their own. Apt to be on yearbook staff, perhaps as an editor. Tend to undertake too much, then somehow get it done. Friendly, but often too absorbed in what they are doing to be sociable or notice much. Live their outer life more with intuition, inner more with feeling.	**INTP** Quiet, reserved, brilliant in exams, especially in theoretical or scientific subjects. Logical to the point of hair-splitting. Interested mainly in ideas, with little liking for parties or small talk. Tend to have very sharply defined interests. Need to choose careers where some strong interest of theirs can be used and useful. Live their outer life more with intuition, inner more with thinking.

Extroverts	Sensing Types: With Thinking	Sensing Types: With Feeling	Intuitives: With Feeling	Intuitives: With Thinking
Judging	**ESTP** Matter-of-fact, do not worry or hurry, enjoy whatever comes along. Tend to like mechanical things and sports, with friends on the side. May be a bit blunt or insensitive. Can do math or science when they see the need. Dislike long explanations. Are best with real things that can be worked, handled, taken apart or put back together. Live their outer life more with sensing, inner more with thinking.	**ESFP** Outgoing, easygoing, accepting, friendly, fond of a good time. Like sports and making things. Know what's going on and join in eagerly. Find remembering facts easier than mastering theories. Are best in situations that need sound common sense and practical ability with people as well as with things. Live their outer life more with sensing, inner more with feeling.	**ENFP** Warmly enthusiastic, high-spirited, ingenious, imaginative. Able to do almost anything that interests them. Quick with a solution for any difficulty and ready to help anyone with a problem. Often rely on their ability to improvise instead of preparing in advance. Can always find compelling reasons for whatever they want. Live their outer life more with intuition, inner more with feeling.	**ENTP** Quick, ingenious, good at many things. Stimulating company, alert and outspoken, argue for fun on either side of a question. Resourceful in solving new and challenging problems, but may neglect routine assignments. Turn to one new interest after another. Can always find logical reasons for whatever they want. Live their outer life more with intuition, inner more with thinking.
Perceptive	**ESTJ** Practical realists, matter-of-fact, with a natural head for business or mechanics. Not interested in subjects they see no use for, but can apply themselves when necessary. Like to organize and run activities. Tend to run things well, especially if they remember to consider other people's feelings and points of view when making their decisions. Live their outer life more with thinking, inner more with sensing.	**ESFJ** Warm-hearted, talkative, popular, conscientious, born cooperators, active committee members. Always doing something nice for someone. Work best with plenty of encouragement and praise. Little interest in abstract thinking or technical subjects. Main interest is in things that directly and visibly affect people's lives. Live their outer life more with feeling, inner more with sensing.	**ENFJ** Responsive and responsible. Feel real concern for what others think and want, and try to handle things with due regard for other people's feelings. Can present a proposal or lead a group discussion with ease and tact. Sociable, popular, active in school affairs, but put time enough on their studies to do good work. Live their outer life more with feeling, inner more with intuition.	**ENTJ** Hearty, frank, able in studies, leaders in activities. Usually good in anything that requires reasoning and intelligent talk, such as public speaking. Are well-informed and keep adding to their fund of knowledge. May sometimes be more positive and confident than their experience in an area warrants. Live their outer life more with thinking, inner more with intuition.

TABLE 4.2 Breakdown of Work Type

Job Function	n	SJ	SP	NT NF
Managers/Supervisors	20	30%	5%	40%25%
Writers	55	15%	4%	56%25%
Editors	11	—	9%	73%18%
Visual Arts	17	24%	12%	18%47%

SOURCE: Hackos (1988), p. RET-18.

GROUP BEHAVIOR CAN BE STUDIED

In addition to being composed of people with individual personalities, groups have personalities of their own. One good way to understand a group's personality is to study it. Attend a group in which you will not be playing a role so that you can be a silent observer of the behavior of others. Draw a picture of the seating arrangement of the group, as shown in Figure 4.1. Each time one person talks to another, draw an arrow from the speaker to the person or persons spoken to. If someone addresses the whole group or no one in particular, draw an arrow outside the group.

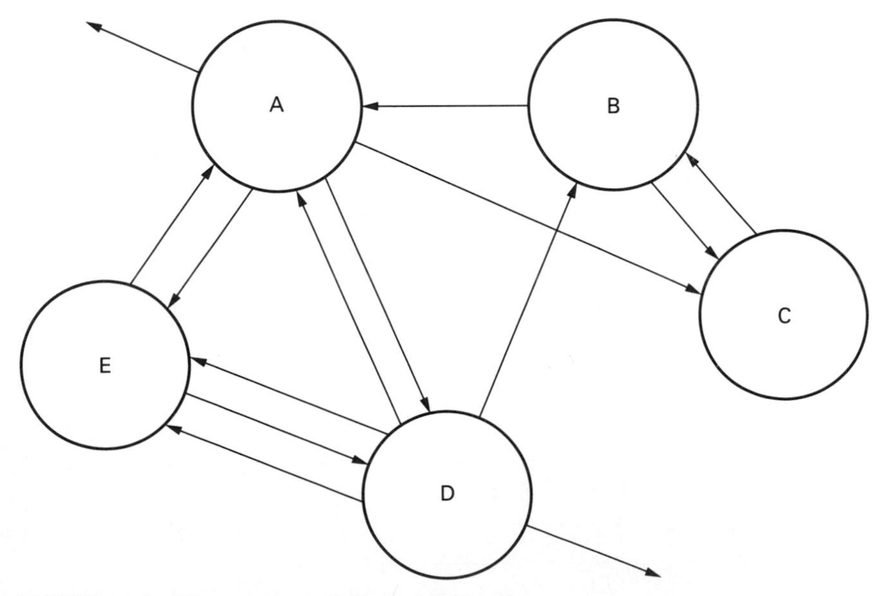

FIGURE 4.1 Observation of Group Dynamics

While observing the group, also pay close attention to the following:

- Observe the posture of the group members. Are some leaning forward, others leaning back? Who is turned toward or away from whom? What are the people doing with their hands?
- Who talks to whom?
- Who doesn't talk much or at all?
- Who talks the most?
- How does position affect interaction with others? Do the people at the corners of the table or room have as much interaction with the people at the ends?
- Is there a designated leader? If so, how would you describe the leader's style?
- If there is no designated leader, who assumes the leadership role? and why?
- If a decision is reached, how does it happen? Were differing opinions expressed? Do people seem committed to the decision?

Interpreting Nonverbal Communication Codes

How do you interpret these behaviors called *nonverbal communication codes*? Schultz (1989), Wilson and Hanna (1986), and Brilhart and Galanes (1989) describe the impact of nonverbal communication codes or "cues" on small group communication by dividing them into several groups. Schultz's grouping (pp. 83-94) includes artifacts, physical environment, proxemics, kinesics, and paralanguage.

Artifacts. *Artifacts* include such cues as dress, hair style, physical features, overall appearance. The choice and style of clothing, jewelry, the way people wear their hair, body shape and size all make impressions on people, who then make judgments about the person's intelligence, confidence, seriousness, persuasiveness, and so forth. These cues are most important in establishing new groups. Once people get to know each other, different cues take precedence.

Physical Environment. The physical attractiveness of the environment also has an impact on communication. This includes room size, temperature, and general appearance. In a classic experiment (reported in Wilson & Hanna, p. 94; Schultz, p. 85), subjects were asked to rate a group of photographs in three settings: a janitor's closet (ugly room), a professor's office (average room), and a beautiful living room (beautiful room). The subjects gave the most positive ratings in the beautiful room, enjoyed the task, and stayed with it longer. Subjects in the janitor's closet complained that they felt irritated and didn't want to finish the task. Schultz concludes that "an unsatisfying environment can cause you to transfer your feelings to the people who are meeting with you" (p. 85). And some people can't understand why having an office window is so important to many writers!

Proxemics. Even more important than the physical characteristics of the room is the arrangement of the people in the group, particularly regarding their distance from one another and choice of seating. These cues are called *proxemics.* Member participation and leadership can be directly influenced by the choices in spacing and positioning. The need to be close for socializing and distant for privacy is influenced by the familiarity with group members, plus other factors such as personality type, people's feelings about the others, sex, race, and age. Studies (reported in Schultz, p. 86) show that cohesive groups occupy less space than noncohesive groups. People of the same age, sex, and race sit closer together than mixed groups. Women sit closer to one another than do men, who not only require more space but also show discomfort when placed in a small room. In contrast, closeness can increase interaction among women. Introverts need more space than extroverts. Higher-status individuals tend to claim more space for themselves (called "territoriality") than do lower-status individuals in a group.

Seating arrangement also sends nonverbal cues. People who choose the head of the table, for instance, are generally the designated leaders or desire to be. Other arrangement factors influence the flow of communication. The emergence of a group leader is largely determined by the person who is in the best position to maintain eye contact with the largest number of members. Howells and Becker (reported in Wilson & Hanna, p. 96; Schultz, p. 88) studied the impact of seating arrangements of five-member groups at a rectangular table. In groups with two on one side and three on the other side of a table, 14 out of 20 leaders were selected from the two-member side. This is attributed to the fact that the two-member side had more eye contact with the three members on the opposite side and that it is more difficult to maintain eye contact with a person sitting next to you than with a person sitting opposite you.

Another experiment by Leavitt (reported in Schultz, pp. 87-88) describes the variables on communication flow in the communication "nets" shown in Figure 4.2: the circle, the wheel, the chain, and the Y. The arrangement of the nets influences the communication flow, the number of messages that can be sent, and the length of time it takes to complete a task. The study showed that "although the wheel is more efficient than the Y, the Y more efficient than the chain, and the circle the least efficient, people are less satisfied when they participate in the wheel, the Y, or the chain patterns" (pp. 87-88). Thus, the circle, which allows for optimum group participation, is the least efficient but most satisfying arrangement for group members. It naturally follows that group work almost always takes more time than individual effort, but the results can be better and more satisfying if the communication net is carefully selected.

Kinesics. *Kinesics* is a term encompassing gesture, posture, movement, and the way people orient their bodies toward or away from others. Research shows that recognized leaders as well as emerging leaders give more positive head nods and gestures than do others

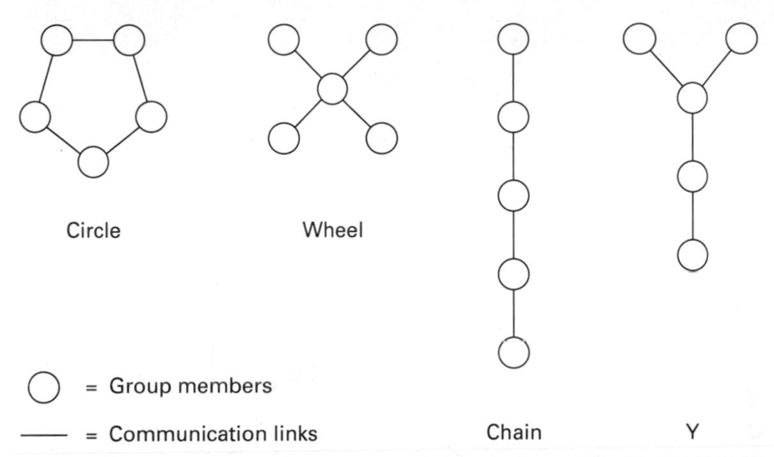

FIGURE 4.2 Communication Nets (Figure from COMMUNICATING IN THE SMALL GROUP by Beatrice Schultz. Copyright © 1989 by Harper & Row, Publishers, Inc. Reprinted by permission of HarperCollins Publishers Inc.)

in a group (Wilson & Hanna, p. 97). Members turn or lean toward those they like and away from those they do not. Members who sit at an angle from the rest of the group may feel excluded or may not want to be included. When members are communicating well, they imitate each other in what is called "postural echoes, body synchrony, or body congruence" (Brilhart & Galanes, p. 147). People tend to follow the leader, as the old game is played. By studying who is imitating whom, you can determine who is perceived as the leader.

Control of the floor, or the right to talk, is also largely regulated by body movement and eye contact. A speaker concluding a point often changes head and eye position or shifts body position. A listener can gain the floor by leaning forward and trying to make eye contact with the leader. Eye contact can also measure a person's status. High-status people receive more eye contact than do low-status people. When eye contact accompanies a message, the message is better received (Schultz, p. 92). Facial features, although often difficult to interpret accurately, are perhaps the most often observed aspect of kinesics. They can communicate interest or boredom, understanding or the lack of it, interpretation and acceptance or rejection, or, in some people, nothing. We've all had the experience of not being able to "read" someone's reaction.

Paralanguage. *Paralanguage* refers to vocal cues, those characteristics of our voice that include inflection, pitch, rate of speech, pauses, and other variables. We often make decisions about a person's education level, ethnic background, and personality on the basis of these vocal cues. People who talk softly are thought to lack persuasiveness because they seem to lack conviction. Others whose vocal qualities change "too often" may be seen as irrational, causing a lack of trust in listeners. Common nonverbal responses—"ums," "uhs," "uh huhs," "mmms"—act as cues to show others that we're listening (Brilhart & Galanes, p. 148).

Recognizing Roles in Groups

Whether we like it or admit it, we all play roles in groups. These are determined partly by personality and partly by the needs of the group. According to the work of Benne and Sheets (reported in Wilson & Hanna), roles can be divided into the following three categories:

- **Task roles** are product-oriented and help a group achieve its goals, as shown in Table 4.3.

TABLE 4.3 Group Task Roles

Roles	Typical Behaviors	Examples
1. Initiator-Contributor	Contributes ideas and suggestions; proposes solutions and decisions; proposes new ideas or states old ones in a novel fashion.	"How about taking a different approach to this chore. Suppose we . . ."
2. Information Seeker	Asks for clarification of comments in terms of their factual adequacy; asks for information or facts relevant to the problem; suggests information is needed before making decisions.	"Wait a minute. What does that mean?" "Does anyone have any data to support this idea?"
3. Information Giver	Offers facts or generalizations that may relate to personal experiences and that are pertinent to the group task.	"I asked Doctor Jones, a specialist in this kind of thing. He said . . ." "An essay in *The New Yorker* reported . . ."
4. Opinion Seeker	Asks for clarification of opinions stated by other members of the group and asks how people in the group feel.	"Does anyone else have an idea on this?" "Can someone clear up what that means?"
5. Opinion Giver	States beliefs or opinions having to do with suggestions made; indicates what the group's attitude should be.	"I think we ought to go with the second plan. It fits the conditions we face in the Concord plant best . . ."
6. Elaborator-Clarifier	Elaborates ideas and other contributions; offers rationales for suggestions; tries to deduce how an idea or suggestion would work if adopted by the group.	"Do you mean he actually said he was guilty? I thought it was merely implied."

(continued)

7. Coordinator	Clarifies the relationships among information, opinions, and ideas, or suggests an integration of the information, opinions, and ideas of subgroups.	"John's opinion squares pretty well with the research Mary reported. Why don't we take that idea and see if . . ."
8. Diagnostician	Indicates what the problems are.	"But you're missing the main thing, I think. The problem is that we can't afford . . ."
9. Orienter-Summarizer	Summarizes what has taken place; points out departures from agreed-upon goals; tries to bring the group back to the central issues; raises questions about where the group is heading.	"Let's take stock of where we are. Helen and John take the position that we should act now. Bill says 'wait.' Rusty isn't sure. Can we set that aside for a moment and come back to it after we . . ."
10. Energizer	Prods the group to action.	"Come on, guys. We've been wasting time. Let's get down to business."
11. Procedure Developer	Handles routine tasks such as seating arrangements, obtaining equipment, and handing out pertinent papers.	"I'll volunteer to see that the forms are printed and distributed." "Look. I can see to it that the tape recorder is there and working. And I'll also run by the church for the chairs."
12. Secretary	Keeps notes on the group's progress.	"I keep great notes. I'll be glad to do that for the group."
13. Evaluator-Critic	Critically analyzes the group's accomplishments according to some set of standards; checks to see that consensus has been reached.	"Look, we said we only had four hundred dollars to spend. What you're proposing will cost at least six hundred dollars. That's a fifty percent override." "Can we all agree, at least, that we must solve the attrition problem—that that is our first priority?"

SOURCE: Wilson & Hanna (1986), p. 145.

- **Maintenance roles** are process-oriented and intended to keep harmony and goodwill in the group. They develop to ease the tensions and disagreements that naturally arise in the team process. As such, these roles, shown in Table 4.4, are essential for group success.

TABLE 4.4 Group Building and Maintenance Roles

Roles	Typical Behaviors	Examples
1. Supporter-Encourager	Praises, agrees with, and accepts the contributions of others; offers warmth, solidarity, and recognition.	"I really like that idea, John." "Priscilla's suggestion is attractive to me. Could we discuss it further?"
2. Harmonizer	Reconciles disagreements, mediates differences, reduces tensions by giving group members a chance to explore their differences.	"I don't think you two are as far apart as you think." "Henry, are you saying . . . ?" "Benson, you seem to be saying . . . Is that what you mean?
3. Tension Reliever	Jokes or in some other way reduces the formality of the situation; relaxes the group members.	"That reminds me—excuse me if this seems unrelated—that reminds me of the one about . . . "
4. Compromiser	Offers to compromise when own ideas are involved in a conflict; uses self-discipline to admit errors so as to maintain group cohesion.	"Looks like our solution is halfway between you and me, John. Can we look at the middle ground?"
5. Gatekeeper	Keeps communication channels open; encourages and facilitates interaction from those members who are usually silent.	"Susan hasn't said anything about this yet. Susan, I know you've been studying the problem. What do you think about . . . ?"
6. Feeling Expresser	Makes explicit the feelings, moods, and relationships in the group; shares own feelings with others.	"Don't we all need a break right now? I'm frustrated and confused and maybe we all are. I'd like to put this out of mind for a while."
7. Standard Setter	Expresses standards for the group to achieve; may apply standards in evaluating the group process.	"In my view, this decision doesn't measure up to our best. We really haven't even set any criteria much less tried to apply them.
8. Follower	Goes along with the movement of the group passively, accepting the ideas of others and sometimes serving as an audience.	"I agree. Yes, I see what you mean. If that's what the group wants to do, I'll go along."

SOURCE: Wilson & Hanna (1986), p. 146.

- **Self-centered roles** arise when an individual's needs become more important than those of the group, often because an individual's needs are not being met. The eight self-centered roles shown in Table 4.5 are counterproductive to the process and the task.

TABLE 4.5 Self-Centered Roles

Roles	Typical Behaviors	Examples
1. Blocker	Interferes with progress by rejecting ideas or taking the negative stand on any and all issues; refuses to cooperate.	"Wait a minute! That's not right! That idea is absurd. If you take that position, I simply can't continue to work with you."
2. Aggressor	Struggles for status by deflating the status of others; boasts; criticizes.	"Wow, that's really swell! You turkeys have botched things again. Your constant bickering is responsible for this mess. Let me tell you how you ought to do it."
3. Deserter	Withdraws in some way; remains indifferent, aloof, sometimes formal; daydreams; wanders from the subject; engages in irrelevant side conversations.	To himself: "Ho-hum. There's nothing in this discussion for me." To the group: "I guess I really don't care what you choose in this case. But on another matter . . ."
4. Dominator	Interrupts and embarks on long monologues; authoritative; tries to monopolize the group's time.	"Bill, you're just off base. What we should do is this. First . . ."
5. Recognition Seeker	Attempts to gain attention in an exaggerated manner; usually boasts about past accomplishments; relates irrelevant personal experiences, usually in an attempt to gain sympathy.	"That was a good thing I just did." "Yesterday I was able to . . ." "If you ask me, I think . . ." "Don't you think I'm right [Don't you think I'm wonderful]?"
6. Confessor	Engages in irrelevant personal catharsis; uses the group to work out own mistakes and feelings.	"I know it's not on the topic exactly, but I'm having a personal problem just like this. Yesterday, Mary and I had a fight about . . ."
7. Playboy	Displays a lack of involvement in the group through inappropriate humor, horseplay, or cynicism.	"Did you hear the one about the cow that swallowed the bottle of ink and mooed indigo?" To the only female in the group: "Hello, sweet baby, let's you and me boogie."
8. Special-Interest Pleader	Acts as the representative for another group; engages in irrelevant behavior.	"My friend Alan runs a company that makes a similar product. How about using his company? We might as well spend our money with people we know."

SOURCE: Wilson & Hanna (1986), p. 148.

Understanding Conflict in Groups

Contrary to popular belief, conflict is not all bad. In fact, a lack of conflict could be very bad, leading to a phenomenon called groupthink, in which everyone agrees, no one points out potential problems, and the group sees itself as an invincible force. The Bay of Pigs disaster is a dramatic example of the problems of groupthink when strategies are not carefully scrutinized and problems fully discussed, even at the expense of a temporary loss of group harmony.

Beebe and Masterson (1989) describe three myths about conflict:

Myth #1 In group discussions, conflict should be avoided at all costs.
Myth #2 All conflict occurs because people do not understand each other.
Myth #3 All conflict can be resolved.
(pp. 209-210)

Contrary to these myths, conflict can be productive, creative, and positive by leading the group to the best solution and by providing ideas and information that challenge group members to think of many options and to analyze advantages and disadvantages of each. Groupthink can be avoided by understanding this healthy aspect of conflict. Conflict can be counterproductive, however, when it leads people away from the task, causes them to lose faith in one another, or, in extreme cases, causes the group to disband. And not all conflict can be resolved.

Because conflict can be productive or counterproductive, it behooves you to recognize which type is occurring so that you can adopt a strategy to manage it. Miller and Steinberg (reported in Beebe & Masterson, pp. 211-213) identify three types of conflict:

- **Pseudo-conflict** occurs when people seem to misunderstand each other and disagree, when, in fact, they agree.
- **Simple conflict** occurs when people understand the issue but have opposing views.
- **Ego-conflict** occurs when someone feels personally attacked by another, and defensive behavior results.

Leadership in Groups

When groups form without a designated leader, one will emerge. Researchers at the University of Minnesota studied this phenomenon to determine who would most likely emerge in a leaderless group. As explained earlier, leaders tend to be those who talk more and give and receive more eye contact. The Minnesota Studies presented some interesting results, divided into two phases (reported in Beebe & Masterson, pp. 252-253):

> **Phase 1. Leaders emerge through a process of elimination.** The first to be eliminated are the quiet ones, the introverts, if you will. Next are the talkative ones perceived as overly aggressive or dogmatic.
> **Phase 2. After half the members are eliminated, the other half vie for the leadership spot.** The group rejects people one by one in a sometimes painful process until one or two remain. Interestingly, the Minnesota Studies found that for groups with two or more men, a male leader was selected, but for groups with only one male, a female was often selected. Task- or product-oriented groups rejected process-oriented leaders, with the reverse being true for process-oriented groups.

SKILLS FOR BECOMING AN EFFECTIVE TEAM MEMBER

By mastering certain skills, you can become an effective team member. While the skills required are many, the following four—storyboarding, improving nonverbal communication, managing conflict, and becoming an effective leader—relate to the previous problems discussed.

Storyboarding

Many people recommend storyboarding as a means to manage the people in group projects and get their full participation. *Storyboarding,* a graphical display of developing text and pictures, can be used for both planning and developing projects. It is a technique often used as a part of managing proposals because it allows for interactive team reviewing of the developing storyboards, which can be posted each day in a conference room dedicated to the project. According to Bacon, this use of storyboarding enhances team commitment because it is easy to see who is doing the work and where the gaps exist. He calls this technique "walking the walls" and states that it not only provides a sense of shared authorship on the developing project, but also helps nurture "better, more capable corporate citizens" (p. 8). For information on storyboarding as a planning tool, see Chapter 2.

Improving Nonverbal Communication

Given all the cues being sent through nonverbal communication channels, you may become anxious about future small group encounters.

Rather than becoming anxious, you can focus on ways to improve the message you send or to increase your chances of being understood, accepted, and valued. Table 4.6 presents the common problems discussed earlier and outlines some strategies for improving nonverbal effectiveness.

TABLE 4.6 Increasing Nonverbal Effectiveness

Problem	Language and/or Suggestions
You find that some members talk too much, others talk too little.	1. Try to regulate the talk patterns by eye contact and head nodding. Avoid eye contact with the talkers; increase your eye contact with those who are silent. 2. Examine the seating arrangement. Maybe the talkers are seated across from each other. In an ongoing group whose members value each other and the group's task, you may be able to approach the more active members and secure their agreement to sit in positions that will minimize the problem. Seat talkative members next to each other if possible. Seat less talkative members across from each other and in central positions. Be careful not to interpret the silent member's nonverbal behavior as a lack of interest. Bormann found that these members were most often experiencing a high level of primary tension, even though they offer the excuse of being disinterested.
You want to be more influential in a new group.	1. Sit at the end of a table if possible. If this is not possible, try to sit on a side where you can address and make eye contact with the majority of the group's members. 2. Make frequent eye contact with members and the group as a whole. 3. Pay close attention to the nonverbal cues. You can provide support, reinforce people, clarify, and the like when you are sensitive to the nonverbal cues others are presenting. These help you to provide both social and task leadership.
What can be done about the ambiguity problem in groups?	1. Check out your perceptions when you can comfortably do so. Use tentative language to let the other person know that you are checking out an inference. You might say either publicly or in private, "I noticed you were [are] staring out the window. You are usually with us. I wonder if there is some problem." 2. Realize as Joseph DeVito has said, "Nonverbal behaviors are normally packaged." What he is saying is that the nonverbal behavior is generally accompanied by other nonverbal behaviors that go along with and support the primary behavior. When you observe someone staring out the window, for example, study the facial expression, posture, and other cues that may help you understand what is happening.

(*continued*)

	3. Try to be aware of the context. Careful attention to the context is always important. Ask yourself, "How does this behavior relate to how the person has been contributing and to what others have been saying?"
You think some members seem to feel excluded.	1. Make eye contact with these people. 2. Ask yourself. "Am I being sensitive to these members' feelings and viewpoints?" If you answer no, or you aren't sure, then try to give more attention to these people.
You are worried about giving off unintentional nonverbal cues in your group. What can be done?	1. Ask a friend (who might also be a member of your group) privately, "I've been wondering how I'm coming off in the group. I want to do my best. How are you experiencing me as a group member?" 2. Self-awareness is the only other answer to this problem. Since you cannot easily observe some of your behaviors, try to be aware of what you are doing as you observe others. Often their reactions will give you a cue to what you are doing.
You think a member of the group is inhibiting communication by certain nonverbal behavior.	1. The behavior might be unintentional. If you think the behavior is unintentional and the member values the group, you may be able to approach the person privately. Be supportive. 2. If the disruptive behavior seems to be intentional, then the group leader may need to talk with the person. A decision about who should talk to the person depends upon who is most likely to be successful. Supportive communication principles should enhance the likelihood of success. 3. The group may need to discuss this problem. Do not take the leader and group member by surprise with such an issue. Check it out with them before the meeting. Be as supportive as possible.

SOURCE: Wilson & Hanna (1986), pp. 100–103.

Managing Conflict

Once you recognize that conflict does and should exist, the next step is to learn strategies for managing conflict. Ewald and MacCallum (1990) show how to make collaborative writing assignments productive by fostering creative tension and minimizing destructive tension. While their suggested activities are intended for the classroom, they emphasize the importance of group maintenance roles and the internal assessment of group interaction. These general principles would serve as well in job-related groups. Table 4.7 shows some strategies for managing the types of conflicts described earlier.

Allen Filley, in his book, *Interpersonal Conflict Resolution* (1975), has made household words of his strategies for managing conflict: win–lose, lose–lose, and win–win.

TABLE 4.7 Suggestions for Managing Conflict

	Pseudo-Conflict	Simple Conflict	Ego-Conflict
Source of conflict	Misunderstanding individuals' perceptions of the problem.	Individual disagreement over which course of action to pursue.	Defense of ego: Individual believes he or she is being attacked personally.
Suggestions for managing conflict	1. Ask for clarification of perceptions. 2. Establish a supportive rather than a defensive climate. 3. Employ active listening: *Stop* *Look* *Listen* *Question* *Paraphrase content* *Paraphrase feelings*	1. Listen and clarify perceptions. 2. Make sure issues are clear to all group members. 3. Use a problem-solving approach to manage differences of opinion. 4. Keep discussion focused on the issues. 5. Use facts rather than opinions for evidence. 6. Look for alternatives or compromise positions. 7. Make the conflict a group concern, rather than an individual concern. 8. Determine which conflicts are the most important to resolve. 9. If possible, postpone the decision while additional research can be conducted. This delay also helps relieve tensions.	1. Let members express their concerns but do not permit personal attacks. 2. Employ active listening. 3. Call for a cooling-off period. 4. Try to keep discussion focused on issues (simple conflict). 5. Encourage parties to be descriptive, rather than evaluative and judgmental. 6. Use a problem-solving approach to manage differences of opinion. 7. Speak slowly and calmly. 8. Develop rules or procedures that create a relationship which allows for the personality difference.

SOURCE: Beebe & Masterson (1989), p. 214.

- **Win–Lose**—in such a conflict, one wins at the expense of the other, who loses. "Forcing," or using power from position is one way in which win-lose strategies prevail. Majority vote is another.
- **Lose–Lose**—compromise is the essence of lose-lose. Compromise means that each person gives up something so that there are no winners. Resentment can occur, especially when compromise results in "pseudo" consensus. Playing down the differences, a technique called "smoothing," can also produce this result.
- **Win–Win**—confrontational problem solving can lead to this result, allowing for consensus-building decisions. Consensus is a stronger result than compromise because it allows everyone to commit to the result. Where problems exist, the focus should be on the goal, not on a particular solution.

Becoming an Effective Leader

The adage says, leaders are born, not made. But it isn't true, as the research amply shows. Leaders develop for different situations and for different needs, as perceived by the group. Some people may feel more comfortable in leadership roles than others, but everyone can profit from learning to be a more effective leader in a group.

Among the important task and maintenance roles that group members play, many of these are leadership roles. So, how can a leader become good at leading? Good leaders

- Understand the goals and the task of the group
- Are good listeners
- Are active participants
- Have good communication skills
- Can clarify issues
- Are open-minded
- Are sensitive and supportive
- Are flexible
- Share rewards and credit with the group

Learning to become effective in each of these areas is something that takes practice and time, but the work can pay off. Since all technical communicators have the opportunity to take on leadership roles in groups, these skills can enhance job success. For those technical communicators intent on management positions, these qualities of effective leadership are vital.

WHAT THE FUTURE HOLDS

The future looks bright for increasing the participation of technical communicators in groups of writers, planners, programmers, and developers. The need to write collaboratively and work harmoniously in groups will only increase. The opportunities for technical communicators to become leaders will also increase.

An issue of *Meeting Management News* (The 3M Meeting Management Inst., 1991) makes some interesting predictions about trends on the horizon as we approach the year 2000. They include the following:

- **Teams:** Organizations are shifting decision making to smaller work units or teams spread through the organization. These teams will rely heavily on groupware and other telecommunications tools.
- **Telecommunications:** Use of telecommunication tools will increase, especially videoconferencing.
- **Intelligent conference rooms:** In the buildings of the future, individual offices will get smaller (because people will be linked electronically), and conference rooms will "resemble large airplane cockpits, with vast control panels for multimedia presentations." Meeting rooms will have video display terminals for individuals with satellite hookup to others in different locations.
- **Meeting management skills:** Since people will be meeting in person in teams and via satellite in videoconferences, meeting management skills will be critical for job success.

While these trends may not become reality by the year 2000, we can safely expect to see many of them in our working lifetimes. Being prepared is the best hedge against obsolescence.

SUMMARY

This chapter has addressed the skills you'll need for career success and job satisfaction. Specifically, this chapter explained that:

- Technical communicators work in groups
- Problems may occur in group process, based on
  - Personality types
  - Group behaviors
  - Nonverbal codes, including
    - Artifacts
    - Physical environment
    - Proxemics
    - Kinesics
    - Paralanguage
- People adopt roles in groups, some helpful, others not, including
  - Task roles
  - Maintenance roles
  - Self-centered roles
- Conflict exists in groups, with some conflict being productive

- Leaders emerge in groups
- You can be an effective team member by mastering the following skills:
  - Using storyboarding techniques for collaborative projects
  - Improving nonverbal communication skills
  - Understanding how to manage conflict
  - Becoming an effective leader

If you're in college, take advantage of the opportunities to learn what you can now about yourself and your communication skills in writing and planning groups. If you're in industry, attend a training seminar or encourage your company to offer one to enhance your understanding of the principles presented in this chapter. Whatever your personality type, you can make a contribution in collaborative projects. You know that you'll find yourself in these situations. You now also know what it takes to succeed.

LIST OF TERMS

artifacts Nonverbal cues given off by such things as dress, hair style, physical features, and overall appearance.

asynchronous type Software for collaborative writing that allows only one person at a time to have access to a document.

boundary spanners Activities of technical communicators as they "translate" information across group lines within the organization and from, and to, an outside audience.

collaborative writing Process of planning, researching, writing, editing, or revising with others.

dialogic model Form of collaborative writing in which the roles are fluid and may shift as the project progresses.

ego-conflict Type of conflict in which someone feels personally attacked by another; this results in defensive behavior.

groupthink Tendency for groups to avoid conflict and reach consensus too quickly without carefully analyzing problems.

horizontal model Form of collaborative writing in which each person takes a segment of the project to write and edit individually.

kinesics Nonverbal cues given off by gestures, posture, movement, and the way we orient our bodies toward and away from others.

lose–lose Strategy in which compromise prevails with the result that there are no winners.

maintenance roles The process-oriented, necessary roles that people assume in groups to keep harmony and goodwill.

Myers-Briggs Type Indicator (MBTI) Profile of personality type developed by Isabel Myers and Katherine Briggs, based on the work of Carl Jung.

nonverbal communication codes Influences on small group communication behavior and perception. Can be grouped into *artifacts, physical environment, proxemics, kinesics,* and *paralanguage.*

paralanguage Nonverbal cues given off by voice inflection, pitch, and pauses.

physical environment Nonverbal cues given off by room size, attractiveness, and temperature.

proxemics Nonverbal cues given off by the arrangement of people in a group.

pseudo-conflict Type of conflict in which people seem to misunderstand each other and disagree, when, in fact, they agree.

self-centered roles Counterproductive roles some members may assume in groups when their individual needs are not being met.

sequential model Form of collaborative writing in which a person writes a draft, then passes it to another who revises it, and passes it to another.

shareware/groupware Terms for the types of computer software that allow for collaborative writing.

simple conflict Type of conflict in which people understand the issue but have opposing views.

small group communication Process by which people interact in groups. Includes nonverbal as well as verbal communication behaviors.

storyboarding Skeletal outline of a project, indicating text and graphics, that can develop as the project is developed. Very commonly used in large projects like proposals.

synchronous type Software for collaborative writing that allows multiple users to have access to the same document at the same time.

task roles Product-oriented, necessary roles people assume in groups that help a group achieve its goals.

vertical (or "stratification") model: Form of collaborative writing in which jobs are assigned by area of specialization of each member.

win–lose Strategy in which one wins at the expense of another, who loses.

win–win Strategy in which confrontational problem solving leads to consensus-building decisions.

REFERENCES

Allen, N., Atkinson, D., Morgan, M., Moore, T., & Snow, C. (1987). What experienced collaborators say about collaborative writing. *Journal of Business and Technical Communication, 1*(2), 70–90.

Bacon, T. R. (1990). Collaboration in a pressure cooker. *Bulletin of the Association for Business Communication, 53*(2), 4–8.

Beebe, S. A., & Masterson, J. T. (1989). *Communicating in small groups: Principles and practices* (3rd. ed.). New York: HarperCollins.

Brilhart, J. K., & Galanes, G. J. (1989). *Effective group discussion* (6th ed.). Dubuque, IA: Wm. C. Brown.

Collins, V. T. (1989). Personality type and collaborative writing. In *Collaborative technical writing: Theory and practice.* Association of Teachers of Technical Writing Anthology Series (pp. 111–116). Urbana, IL: ATTW.

Couture B., & Rymer, J. (1989). Interactive writing on the job: Definitions and implications of "collaboration." In M. Kogen (Ed.) *Writing in the business professions* (pp. 73–93). Urbana, IL: National Council of Teachers of English and The Association for Business Communication.

Cross, J. A. (1990). A Bakhtinian exploration of factors affecting the collaborative writing of an executive letter of an annual report. *Research in the Teaching of English, 24*(2), 173–203.

Debs, M. B. (1991). Recent research on collaborative writing in industry. *Technical Communication,* Fourth Quarter, 476–484.

Doheny-Farina, S. (1986). Writing in an emerging organization: An enthnographic study. *Written Communication, 3*(2), 158–185.

Ede, L., & Lunsford, A. (1985). Research into collaborative writing. *Technical Communication*, Fourth Quarter, 69–70.

Ede, L., & Lunsford, A. (1990). *Singular texts/plural authors: Perspectives on collaborative writing.* Carbondale: Southern Illinois University Press.

Elwart-Keys, M., & Horton, M. (1990). Collaboration in the capture lab: Computer support for group writing. *Bulletin of the Association for Business Communication, 53*(2), 38–44.

Ewald, H. R., & MacCallum, V. (1990). Promoting creative tension within collaborative writing groups. *Bulletin of the Association for Business Communication, 53*(2), 23–26.

Faigley, L. & Miller, T. P. (1982). What we learn from writing on the job. *College English, 44,* 557–569.

Farkas, D. K. (1991). Collaborative writing, software development, and the universe of collaborative activity. In M. M. Lay & W. M. Karis (Eds.) *Collaborative writing in industry: Investigations in theory and practice* (pp. 13–30). Amityville, NY: Baywood.

Filley, A. (1975). *Interpersonal conflict resolution.* Glenview, IL: Scott, Foresman.

Fischer, S. (1991). Stop, look, & listen–what you heard may not be what was said. How to overcome messages that stifle or misdirect a project. In *Proceedings of 38th ITCC* [14-17 April]. (pp. ET 113–115). New York: Society for Technical Communication.

Hackos, J. (1988). Personality type in technical communication. In *Proceedings of 35th ITCC* [10–13 May]. (pp. RET 16–18). Philadelphia: Society for Technical Communication.

Halpern, J. W. (1985). An electronic odyssey. In L. Odell & D. Goswami (Eds.). *Writing in nonacademic settings* (pp. 157–202). New York: The Guilford Press.

Harrison, T. M. & Debs, M. B. (1988). Conceptualizing the organizational role of technical communicators: A systems approach, *Journal of Business and Technical Communication, 2*(2), 5–21.

Hickman, D. E. (1991). Neuro-Linguistic Programming tools for collaborative writers. In M. M. Lay & W. M. Karis (Eds.), *Collaborative writing in industry: Investigations in theory and practice* (pp. 211–223). Amityville, NY: Baywood.

Hirokawa, R. Y., & Gouran, D. S. (1989). Facilitation of group communication: A critique of prior research and an agenda for future research. *Management Communication Quarterly, 3*(1), 71–92.

Keller, C., & Knapp, S. (1991, February 7). Guest lecture presented to Management Communication class, Southern Tech, Marietta, GA.

Lay, M. (1989). Interpersonal conflict in collaborative writing: What we can learn from gender studies. *Journal of Business and Technical Communication, 3*(2), 5–28.

Louth, R., & Scott, A. M. (Eds.). (1989). *Collaborative technical writing: Theory and practice.* Association of Teachers of Technical Writing Anthology Series. Urbana, IL: ATTW.

Lunsford, A., & Ede, L. (1986). Why write . . . together: A research update. *Rhetoric Review, 5*(1), 71–77.

Lunsford, A., & Ede, L. (1990). Rhetoric in a new key: Women and collaboration. *Rhetoric Review,* 8 (2), 234–241.

Myers, I. B. (1980). *Introduction to type.* (3rd ed.) Palo Alto, CA: Consulting Psychologists Press.

Raymond J., & Yee, C. (1990). The collaborative process in professional ethics. *IEEE Transactions on Professional Communication 33*(2), 77–81.

Samson, D. C. (1989). Collaborative writing of technical proposals. In R. Louth & A. M. Scott (Eds.), *Collaborative technical writing: Theory and practice.* Association of Teachers of Technical Writing Anthology Series (pp. 61–69). Urbana, IL: ATTW.

Schultz, B. G. (1989). *Communicating in the small group: Theory and practice.* New York: Harper & Row.

Shirk, H., (Ed.) (1990). Special issue: Collaborative writing in business communication. *Bulletin of the Association for Business Communication, 53(2).*

Sides, C. H. (1988). Carl Jung and documentation life cycles. In *Proceedings of 35th ITCC* [10-13 May]. (pp. RET 94–95). Philadelphia: Society for Technical Communication.

Sides, C. H. (1989). What does Jung have to do with technical communication? *Technical Communication,* Second Quarter, 119–126.

Sides, C. H. (1991). Collaboration in a hardware technical writing group: A real-world laboratory. *Bulletin of the Association for Business Communication, 54(2),* 11-15.

Smit, D. W. (1989). Some difficulties with collaborative learning. *Journal of Advanced Composition, 9,* 45–58.

Stratton, C. R. (1989). Collaborative writing in the workplace. *IEEE Transactions on Professional Communication, 32(3),* 178–182.

The 3M Meeting Management Institute. (1991, April). The year 2000: Expect meetings to change, not decline. *Meeting Management News,* 1–3, 8.

Wilson, G. L., & Hanna, M. S. (1986). *Groups in context: Leadership and participation in small groups.* New York: Random House.

PART II

TECHNIQUES FOR DESIGNING INFORMATION

CHAPTER 5

A WAY WITH WORDS—PRESENTING INFORMATION VERBALLY

Saul Carliner

Technical
communicators
present many
types of information

INTRODUCTION

Earlier chapters have told you how to develop a plan for information. Now, how do you turn it into prose? For example, how do you write the introduction? For that matter, what type of introduction do you write? A lead? A grabber? How do you write the body? For example, how do you write a procedure? How do you write a reference entry? What type of reference entry do you write? A command description? A policy? . . .

Answering this seemingly endless list of questions becomes easier as you develop a repertoire of presentation techniques. From this repertoire, you "mix and match" techniques as needed to present specific types of information and integrate them cohesively into a document. This chapter focuses on verbal presentation techniques; that is, presenting information with words. (Later chapters focus on other presentation techniques.) Specifically, this chapter explains the following:

- Considerations for choosing a presentation format or technique
- Several formats for presenting information verbally including:
  - Formats for introducing passages
  - Formats for presenting information by
    - Instructing
    - Persuading
    - Referencing
  - Formats for summarizing passages
  - Techniques for writing generically (a cost-cutting measure)

This chapter explains what each verbal presentation technique is as well as when and how to use it. This chapter also provides examples of most presentation techniques. Where you might be confused about a presentation technique, this chapter provides an example of what you shouldn't do. This chapter concludes with a discussion of two key issues you should consider when writing and a detailed example explaining how to select verbal presentation techniques.

CONSIDERATIONS FOR CHOOSING A PRESENTATION FORMAT OR TECHNIQUE

According to Edmond Weiss (1991):

> There are two broadly different ways to write a document. The first is to compose it, crafting the sentences and paragraphs while they are

> being written, as would a writer working on a script. The second is to *engineer* it, preparing a series of increasingly finer specifications until, at last, a document drops out. (p. 40)

Weiss alludes to the two components of writing: artistry and craftsmanship. Artistry pertains to the ability to conceive ideas and is stereotypical of the writing process. Craftsmanship pertains to the ability to express ideas and is typical of the engineering process that Weiss describes. Any good document represents a blend of artistry and craftsmanship. This chapter focuses on craftsmanship.

The craft of technical communication is expressing information in such a way that people can easily make use of it. This craft is the ability to write precise, accessible, and "simple" (as in not elaborate) prose. In this chapter, craftsmanship refers to the ability to choose the most appropriate technique for presenting information.

In a few instances, the technique for presenting information involves following a format that has already been chosen for you. A format is a plan for the organization and arrangement of information. For example, research reports generally follow a certain outline and present similar types of material. More often, your document contains several types of information and you must use several formats, each specifically chosen according to the information you are presenting.

Suppose you are writing a hypertext stack (a type of online document) that describes employee benefits. Although your stack primarily consists of several reference entries, you might need to include definitions of some key terms, like "coordination of benefits" and "dependent survivors." These definitions need to be prepared in a different format from the reference entries.

Or, suppose you are writing a video that presents the procedure for assembling computer disk drives. Although the information is primarily procedural, you also need to include an introduction that motivates viewers to pay close attention because many of the viewers are already familiar with the procedures but are not following them.

In each of these instances, you address some sort of communications problem. Not surprisingly, Linda Flower (1978) calls technical writing a problem-solving discipline. Presenting information is part of the problem-solving process. It involves more than merely writing information and ensuring that it is grammatically correct. Specifically, presenting information involves these steps:

- Assessing the type of information to be presented. For example, you need to determine whether the information you are presenting is a procedure or an appeal to people to buy something.
- Finding a presentation format that addresses your goals and readers' needs. For example, you might present the procedure as a series

of steps that readers can easily follow or relate your sales appeal to a need that readers have.

The following sections present several of the most commonly used formats and techniques for presenting information verbally. The list of formats and techniques is not exhaustive but provides the tools needed for handling the most common assignments in technical communication. Some of these formats come directly from research, whileothers derive from a combination of research and on-the-job experience. Chapter 6 continues this discussion with several common formats for presenting information visually.

INTRODUCING PASSAGES

Introductions

What an Introduction Is. An *introduction* is a paragraph or group of paragraphs at the beginning of a passage (a generic term that refers to related information, whether that information be a group of paragraphs, a section, a chapter, or an entire document). An introduction

- Grabs and holds onto readers' attention
- Relates information to readers' needs
- Presents an overview of the upcoming passage

Many terms exist for introductions. The following terms refer to different types of introductions, each adapted for its own particular use:

- *Grabber* is used in training and advertising to gain the attention of the audience.
- *Lead* is used in journalism to begin a passage with the most important points.
- *Opening* is used in technical writing and training to give an overview of an upcoming passage and relate it to readers' needs.

How Introductions Benefit Readers. Readers don't judge a book just by its cover; they also judge it by their first few moments of reading. These first few moments almost always happen in the introduction, so you should invest extra energy in crafting introductions.

The audience for technical information is not necessarily captive (Brusaw, Alred, & Oliu, 1976, p. 318). No wonder that "some experts say you win or lose your reader in the first five lines" (Frank, 1985, p. 105). As explained in Chapters 1 and 2, readers don't just read passages on their own. You need to convince them to do so. Introductions help you do that by:

- Relating information to readers' needs
- Relating new information to information readers already know, giving them something to "hook" the new information onto
- Showing readers why the information is important
- Providing an overview of the information to be presented

In short, introductions make information accessible to readers because they tell readers why they should read a passage.

When to Use Introductions. Use an introduction in any passage intended to be read from start to finish. The passage can be a section of a document or the entire document. Some documents might have several introductions because individual sections are intended to be read from start to finish, even if the entire document isn't. Specifically, use introductions with these types of information:

Type of Information	Why You Should Use Introductions
Sales information	According to Schroello (1990), few readers (less than 5%) respond to sales literature. Introductions (called grabbers) might help increase the yield by causing readers to pay attention.
Instructional	"To the degree that we can . . . conceptually anchor [new material] to what [readers] know, that new material will be meaningful" (Jonassen, 1982, p. 256). Similarly, readers must also see the value of an instructional unit before they actively participate in it. Introductions perform both jobs: anchoring new material to existing knowledge and showing readers the value of information.
News (such as articles for newsletters and press releases)	Because more news is available than people have time to use, introductions (called leads) help readers get the gist of a news story and determine whether they need to read further.
Reports and scholarly and professional articles	With the amount of information doubling every three years (Braun, 1991), readers need some tool to help them determine which information they should read.

Do not use introductions as just described for sections of reference material. Readers already know what they're looking for and usually ignore the introduction. For example, do not use an introduction to a parts catalog. Instead, use a preface, which tells readers *how* to use the catalog.

How to Write Introductions. The type of introduction you use depends on the type of information you are presenting. Following are suggestions on the types of introductions to use with particular types of materials. Each refers you to a separate section in this chapter for more detailed instructions:

Type of Information	Type of Introduction You Should Use
Instructional	Opening (and, sometimes, grabber)
Memos	Lead
News (articles for newsletters and press releases)	Lead
Reports and scholarly and professional articles	Opening (and, sometimes, grabber)
Sales	Grabber
References (such as command and medical reference manuals)	Preface
Other documents meant to be read start to finish	Opening

Openings

What an Opening Is. An *opening* is a paragraph or group of paragraphs at the beginning of a passage that provides an overview of upcoming information and relates that information to readers' needs.

How Openings Benefit Readers. In addition to the benefits mentioned earlier, openings tell readers the main points of the upcoming passage. Knowing this information in advance, readers can distinguish main points from subordinate ones in the body of a text and know which information they should pay most attention to. They can also determine whether the information is intended for them.

When to Use Openings. Use openings in these instances:

Type of Information	Why You Should Use Openings
Instructional materials	To set realistic expectations and to tell readers what they should pay attention to
Reports, scholarly and professional articles, other information intended to be read from beginning to end	To tell readers why the article or report is noteworthy (what gap in information it fills), to tell readers what they should specifically pay attention to, and to provide the background or historical information needed to understand the rest of the article or report

How to Write Openings. Openings should have three parts:

1. Grabber, intended to gain readers' attention. Because grabbers are a unique feature, they are described in another section in this chapter.

2. Motivator, which relates the information in the passage to information that readers already know and explains how the information addresses their needs.
3. Overview of the passage. The overview is presented in one of two ways:
   - In instructional materials, the overview consists of the instructional objectives presented in the order they are discussed.
   - In other materials, the overview lists the major topics to be addressed and shows which points are main ones and which are subordinate ones.

Example of What You Should Do

CHAPTER THREE: USING THE WORD PROCESSOR

Previous chapters gave an overview of the word processor and explained how to install it. This chapter explains how to use the word processor.

Objectives

After reading this chapter, you should be able to

- Enter text
- Copy, move, and delete text
- Set margins and tabs
- Emphasize text using underlining, and bold and italic type
- Save text

Examples illustrate how to perform these tasks.

Grabbers

What a Grabber Is. A *grabber* is a sentence, paragraph, or group of paragraphs at the beginning of a passage that gains—or grabs—and holds readers' attention.

How Grabbers Benefit Readers. According to Gagné (1985, p. 304), before you can inform or educate readers you need their attention. Grabbers gain attention.

In doing so, grabbers develop readers' interest in a passage that they might otherwise skip and make seemingly forbidding information accessible.

When to Use Grabbers. Use grabbers for any material that might be overlooked or for which the motivation to read is low. Specifically, use grabbers when introducing the following types of information:

Type of Information	Why You Should Use Grabbers
Instructional and informational materials that readers are required to read	Motivation and attention levels are low. A clever grabber can raise these levels.
Sales information	People encounter so much sales information that they often toss it before reading it. A clever grabber can catch readers' eyes before they discard the information.

How to Write Grabbers. Writing grabbers provides you with one of the most creative opportunities in technical communication. Anything can serve as a grabber if it:

- Directly relates to the subject matter. By relating to the subject matter, grabbers can focus attention on your topic.
- Is motivational to readers. What motivates one group of readers might not motivate another group. For example, a group of executives who receive stock options might be motivated by a grabber that relates an upcoming passage to the health of the company's stock. A group of customers would not be motivated by that concern.

"Creative" motivational material is often a matter of taste. What might seem appropriate and tasteful to you might not be appropriate and tasteful to your readers. When writing a grabber, then, consider how readers might interpret it. If you are not sure, ask a few people who are representative of the intended readers to tell you how they interpret the grabber.

Although grabbers are unique by nature, they fall into several general categories. The type of grabber you use depends on the type of information you are writing.

Type of Grabber	Suggested Use	Example of a Grabber
Focus on the benefits of the information (Price, 1984, p. 69)	Training materials, user manuals	Here's an easier way to complete the audit:
Go from familiar to unfamiliar information	Passages with new concepts	As you become more familiar with an environment after living in and participating in it, so ethnographers become familiar with an environment by living in and participating in it. The only difference is, they report their experiences.

(*continued*)

Type of Grabber	Suggested Use	Example of a Grabber
Present a catchy phrase (be careful, what's catchy to one person might not bc catchy to another)	Passages readers might otherwise miss	Where can you save 50% on office furniture? At our next inventory clearance.
Present a visual to "motivate readers to read and act" (Horton, 1991, p. 196)	Introductory material that's visual or leads to a visual response, such as before and after pictures	Begin with a photo of contented person. Copy would read: This could be you after six months of using our time management system.

Example of What You Should Not Do

> **WIN A MILLION BUCKS.**
>
> Actually, no, but have you ever considered the value of office safety?

This is a poor example of a grabber. Although it gains readers' attention, this grabber focuses attention on winning a million bucks rather than on the main subject, office safety.

Leads

What a Lead Is. A *lead* is a paragraph at the beginning of a passage that introduces all the main points of the passage.

How Leads Benefit Readers. Leads benefit readers by giving them an immediate overview. Readers can then determine whether they need to read the entire passage.

When to Use Leads. Use leads to begin passages that compete with other passages for readers' attention, such as

- News articles
- Press releases
- Memos

How to Write Leads. A lead provides all the essential information about a passage, such as

- Who is involved
- What happened
- When it happened

- Where it happened
- Why it happened
- How it happened

Journalists call these the five Ws and H.

The challenge of writing leads is being brief. The ideal lead is no more than 30–50 words, like the leads in most news articles.

In a memo, you also include a call to action or purpose statement in the lead. This tells readers what you expect them to do after reading the memo.

Examples of What You Should Do. The following sentence is from a news story for an employee newsletter:

> New healthcare benefits announced today provide employees with greater flexibility because they can choose coverage from a Health Maintenance Organization (HMO) or a private provider.

Note the five Ws and H:

Who	Employees
What	Choose healthcare benefits
When	Announced today
Where	Everywhere in the company (implied)
Why	Greater flexibility
How	Choose coverage

From a press release:

> Marion Anderson, president of Widgit Corporation, announces the appointment of Charles Quigg as Vice-President for Technical Development, effective January 27, 1993.

Note the five Ws and H:

Who	Charles Quigg
What	New Vice-President
When	January 27, 1993
Where	Widgit Corporation

Why Typically, people are promoted because they do a good job, so the writer assumes that readers already know "why."

How Through this announcement

The next sentences are from a memo:

> Clean It, the janitorial service we currently use, provides superior service at the best price. Please reward Clean It with a two–year extension of its contract.

Note the five Ws and H:

Who	Clean It (a company)
What	Reward
When	Now
Where	At our building (implied)
Why	Superior service at the best price
How	With a two-year extension of the contract

Because this is a memo, it includes a call to action—"reward Clean It with a two-year extension of its contract."

Prefaces

What a Preface Is. "A *preface* is a statement... about the purpose, background, or scope" of information (Brusaw, Alred, & Oliu, 1976, p. 180).

A preface is usually a section at the beginning of a document that explains the following:

- What the document is about
- Who its intended readers are
- What readers should already know before reading the document (if appropriate)
- How readers should use the document
- What readers need to have to use the document
- Other useful information about the document

A document can contain both a preface and an introduction. When a document contains both, the subject of the introduction is the information in the document, and the subject of the preface is the document itself. Readers would read a preface to determine whether a document meets their needs. For example, a student might read the preface of a reference book to determine if it might provide the information needed for a class project. An instructor might read the preface of a new text to determine whether to assign the book for class.

How a Preface Benefits Readers. A preface benefits readers by telling them what they need to know to use a document effectively and efficiently.

When to Include a Preface. Include a preface in the following situations:

- With manuals, when readers need to be aware of special instructions (such as prerequisite knowledge and availability of equipment)
- With software packages, when readers need to be made aware of requirements for using the software (such as the equipment and other software needed)

How to Write a Preface. A preface consists of several clearly marked sections, including the following. Note that the headings provided for these sections are generic; you can tailor them to the needs of your document.

Purpose — Explains what the document is about. Be brief; explain the purpose in 50 words or less. For example:

> This *Problem Determination Guide* provides step-by-step procedures for identifying and solving problems with your computer.

Audience — Explains who the intended readers are. Be as specific as possible, including job titles and education levels where appropriate. For example:

> This *Problem Determination Guide* is intended for system operators who have experience with starting, stopping, and interrupting jobs.

Note the mention of both a job title and job skills in this example. In some instances, the reader might be performing the skills needed to use the guide but might have a different job title.

Prerequisite knowledge — Explains what readers should already know before reading the document (if such knowledge is needed). Explains the prerequisite knowledge in terms of skills readers should have rather than as other documents readers should read. Lists the other documents as sources for learning those skills. Readers can determine whether they need to read the prerequisite documents. For example:

> Before using this *Problem Determination Guide*, you should be able to:

- Start and stop the system
- Start and stop application programs
- Operate the printer
- Enter JCL commands

This information is presented in *The System Operator's Guide.*

How this document is organized	Tells readers where to find information in the document. For example: This Guide begins with an initial procedure to determine the exact nature of the problem. It continues with several specific procedures, each exploring a different type of problem, such as a hardware problem or a printing problem.
Conventions used in this document or instructions for using this document (both titles are used)	Tells readers how to use the document if the instructions are not self-evident. This section is especially useful for online documents, whose structures are not readily apparent to readers. If instructions and conventions are not self-evident, however, you should re-examine their use. As one technical writer commented, "Manuals should be like rental cars. You should be able to use them without *another* instruction manual."
What readers need when using the document	Varies, depending on the medium used for the document. ■ Online documents should have a "Machine Requirements" section, identifying the hardware and software needed. Print this information on the outside of the software package rather than in the software; readers need this information before they install the online document. ■ Instructional materials (videos, workbooks, and computer-based tutorials) should have a list of all documents and other materials in the course so instructors can make sure they have all the materials needed. For example: **Special Instructions:** You need the following equipment when taking this lesson: □ IBM-compatible computer □ DOS 5.0 □ Windows™ [Windows is a trademark of Microsoft Corporation.] □ Videodisc player □ VGA monitor

Where you place the preface varies, depending on the document you are producing. Following are some suggestions:

Type of Document	Where You Should Place the Preface
Printed documents	At the beginning of the document
Online documents	In one of these places: ■ For self-contained documents, on the printed package accompanying the document ■ For online documents that are part of a series of online documents, within the document; provide readers with an option called, "About this Document"
Videos	On the video jacket

Although the preface contains information essential to readers' success in using a document, experience shows that few readers actually read it. One of your challenges, then, is finding ways to encourage readers to notice the preface. Some ways to accomplish this are:

- Printing the preface on the cover of the manual or of the package holding the document so readers can't miss it
- Using eye-catching graphical devices, such as those described in Chapters 6 and 7
- Repeating vital information from the preface elsewhere in the document (in case they skipped the preface)

Example of What You Should Do. Following is an example of a preface for a training video for a new camera. The headings of the sections have been tailored to this video and several sections have been omitted. This preface would be printed on the video jacket.

About This Video: This video explains how to operate your new ABC Deluxe 35mm camera.

What You Should Know: You need no previous photography experience to understand this video–or operate this camera.

How to Get the Most from This Video: First, watch it from beginning to end. Then, to master specific skills, watch specific sections, such as the section on setting the shutter speed and the section on choosing film. The label on the video case tells the counter locations of these sections in the video.

You might also want to have your camera nearby so you can practice skills as the narrator demonstrates them.

PRESENTING SPECIFIC TYPES OF INFORMATION

The presentation of information is the heart of any technical document. Sometimes, this part of the document is called the *body.*

Types of Information That Technical Communicators Present

Technical communicators are best known for presenting procedures and reference information, especially for computer software. Technical communicators also present other types of information, including:

- Sales proposals for large contracts. These proposals explain the benefits of choosing a certain contractor and describe how that contractor can meet the customer's needs.
- Collaborative reports with scientists describing scientists' research to the scientific community.
- Newsletters that keep employees informed about recent developments in their organizations.
- Catalogs that provide detailed information about products for potential customers.

Formats Available for Presenting Information. Several formats are available for presenting information. You can think about these formats in categories. Many schemes are available for categorizing formats. Boiarsky (1991) suggests three categories of formats: expressive, persuasive, and referential. Here, they are renamed as follows:

- Formats for instructing (expressive category). These formats "teach" readers about your technical subject. Instructional passages can teach the following:
    - Concepts are sets of characteristics uniquely common to a set of similar things, beings, or abstract ideas. (If this definition is a little fuzzy, read on. I clarify it in a little bit.) Technical communicators use several techniques to present concepts:
        - Explanations define a concept and clarify it for readers.
        - Analogies compare new concepts to ones that readers are familiar with.
        - Examples can be used to create a special technique for teaching concepts.
    - Procedures explain how to perform a task.
- Formats for persuading (persuasive category). These formats encourage readers to take a particular action or point of view:

- Sales information persuades readers to purchase a product or service.
- Scholarly and professional articles persuade readers to a particular view. Often, scholarly articles appear to present facts. In most instances, however, the authors present facts to support a particular theory.

- Formats for referencing (referential category). These formats provide factual information to readers:
  - Reference entries, which describe elements of a system.
  - Glossaries, which are special types of references for presenting terms.

More important than the names for these categories and the formats within them are the expectations that readers have when reading information in each of these formats. These formats can also be called *genres.* According to Foss (1989, pp. 111–112), a genre of information is created when different communicators address similar situations in a similar way. Specifically, communicators writing in a particular genre include similar types of content and write in a similar style. For example, a reference entry for a computer software command always includes an example. A procedure has numbered steps.

As a technical communicator you need to be able to understand which genre or format you should use in a situation and proficiently write in that format.

The next several sections describe the formats for presenting many types of information.

Formats for Instructing

What Instructional Formats Are. Instructional formats teach readers about a technical subject.

When Technical Communicators Need to Instruct. Use instructional formats when your readers have little or no familiarity with the subject.

Formats Available for Instructing. Include the following:

- Concepts or ideas. Use the following tools to present concepts:
  - Explanations
  - Analogies
  - Examples
- Procedures, which explain how to perform a task.

Concepts

What Concepts Are. People commonly think of a concept as "a thought, or an idea" (Brusaw, Alred, & Oliu, 1976, p. 108). But in instruction, a concept is more than just a thought or idea. A concept is the set of characteristics that are common to a set of similar things, beings, or abstract ideas. A concept can be a dog, a religion, a computer, or an approach to problem solving.

According to Becker (1986, p. 185), a concept "can be defined by . . . properties . . . common to the concept instances and not common to non-instances." That is, a concept has certain characteristics. For example, dogs have certain physical characteristics that distinguish them from cats and other types of animals. Poodles have certain physical characteristics that distinguish them from other types of dogs.

How Presenting Concepts Benefit Readers. A clear presentation of a concept benefits readers in many ways. At the most basic level, the presentation tells readers what they are learning.

The presentation also provides readers with prerequisite information (information they need to learn before they can learn the lesson at hand.) For example, lab technicians need to learn how AIDS is transmitted before working with blood samples.

Finally, the presentation shows readers how to distinguish that concept from similar ones. Later on, when readers are on their own, they can make the distinctions without assistance. For example, suppose your employee handbook explains which information should be classified as confidential. When employees encounter classified information on the job, they will know how to properly handle it.

When to Explain Concepts. Explain concepts when you are presenting these types of information:

Type of Information	Why You Should Explain Concepts	Example of Situation
Abstract material	Helps readers understand the material	In writing a manual explaining how to improve system throughput (the speed at which information flows through a computer), you must first explain the way information flows through the computer, an abstract concept.
Procedure involving judgment	Provides readers with the tools needed to make a knowledgeable assessment	In writing an expert system for assessing damage to parts being manufactured, you must first explain the different types of damage and their effect on the entire line so operators on the manufacturing line can properly use the recommendations of this system.

When You Should Not Explain Concepts. You do not always need to explain concepts. If understanding a concept is not essential to using the information, don't explain it. (In fact, presenting the concept in these instances distracts readers from the main point.)

One of the most obvious examples of this is cars. People can drive a car without understanding how the engine works. Similarly, you do not need to explain what DOS is when explaining how to install a software application for a personal computer. Readers need to know only the commands for installing DOS.

How to Present Concepts. Although no formula exists for presenting concepts, you can use one or more of the following presentation tools:

- Explanations
- Analogies
- Examples

The next sections describe these tools.

Explanations

What an Explanation Is. An *explanation* "sets the stage" (Brusaw, Alred, & Oliu, 1976, p. 163). It is a sentence, paragraph, or group of paragraphs that defines a concept and presents associated rules and definitions.

How Explanations Benefit Readers. Explanations benefit readers by telling them exactly what you mean by a particular term or idea. Providing clear explanations ensures that you and your readers have a similar understanding of the term or idea.

When to Use Explanations. Explain any concept that is new or unfamiliar to your readers.

For example, some textbooks for undergraduate students use the terms "heuristic" and "empirical," but many readers may not be familiar with these terms. It would be appropriate, then, to explain any terms that may be unfamiliar to the intended audience.

How to Write Explanations. Many techniques are available for writing explanations. Space permits the opportunity to present only one.

According to Becker (1986), a good explanation tells readers how to generalize a concept to all its instances. That is, if you explain what a modem is, readers should be able to use that explanation to identify a modem as a modem every time they see one.

An explanation, then, must include all of the main properties of the concept you are explaining. You add to that explanation examples and analogies to ensure that readers understand the explanation.

Example of What You Should Do. Following is an explanation of a "monthly activity report":

> A monthly activity report is intended for your manager and includes a list of all major activities you performed during the month. Major activities include progress on assigned projects, regularly assigned duties, and special projects handled during the month. Monthly activity reports should also include an account of all training received during the month. Your manager compiles information from each employee's monthly activity report to develop a monthly activity report for the entire department. Monthly activity reports are due the second business day of the following month.

Notice that the explanation presents the key features of the monthly activity report as well as its intended audience and deadline.

Example of What You Should Not Do. Following is a poor explanation of the term "monthly activity report":

> A report you turn in to your boss at the end of the month explaining your activities for the month.

Although the explanation tells you the main content of the report and its primary audience, the explanation fails to clarify:

- What the term "activities" means
- How your boss uses the report
- The exact date that the report is due ("end of the month" is not as precise as "second business day of the following month")

Analogies

What an Analogy Is. An *analogy* is a comparison that relates an unfamiliar concept to a familiar one by showing what the similarity is. An analogy can provide "a shortcut means of communication if it is used with care" (Brusaw, Alred, & Oliu, 1976, p. 37).

How Analogies Benefit Readers. As noted in Chapters 1 and 2, readers best understand new information when you present it in terms of something they already know because readers "attach" the new information to information already stored in memory. As a result, readers are more likely to retain and use the information. The better you understand your readers, the more relevant the analogies you can choose.

When to Use Analogies. Use analogies when writing the following types of materials:

Type of Information	Why You Should Use Analogies
Instructional	To help readers learn new concepts
Sales literature	To show potential customers how your product might be useful to them
Reference	To help readers understand the concepts

How to Write Analogies. According to Newby and Stepich (1990), you write analogies using the following formula:

Subject	Connector	Analog	Ground
SOMETHING NEW	is like	SOMETHING OLD	QUALIFIER
Online information	is like	a book	in that both contain information.

Note the "ground" in the formula. The ground explains how the two concepts are similar. The use of a ground distinguishes analogies in technical communication from those in fiction, which merely state the comparison. As Newby and Stepich also suggest, you should use a visual aid to reinforce analogies when appropriate.

Examples of What You Should Do

> A computer file is like a file folder in that both hold information.

(subject = computer file, analog = file folder,
ground = both hold information)

> A person's short-term memory is like a computer's random access memory in that it is an intermediate stop in the flow of information to long-term storage and has a limited capacity.

(subject = person's short-term memory,
analog = computer's random access memory,
ground = an intermediate stop . . . capacity)

> A professional organization is like a trade guild in that it is a group of people with similar job skills.

subject = professional organization, analog = trade guild,
ground = group of people with similar job skills)

Examples of What You Should Not Do

A computer file is like a file folder.

(The ground is missing.)

A professional organization is a group of people with similar job skills.

(The analog is missing.)

Examples and Non-Examples

What Examples and Non-Examples Are. In instruction, an *example* is an instance of a concept. That's a little abstract, so consider this in more concrete terms. A sheepdog is an example of the concept "dog." A supercomputer is an example of the concept "computer." Health insurance is an example of the concept "employee benefits."

Closely related to examples are non-examples. According to Horn (1983), a *non-example* is something that might be confused with the concept because it is similar to the concept. The non-example, however, does not have all the features of the concept and, therefore, cannot be used as an example. Consider that frozen yogurt is a non-example of ice cream. Aspartame is a non-example of sugar.

How Examples and Non-Examples Benefit Readers. Examples benefit readers by clarifying concepts and expressing them in concrete terms. Examples provide readers with an image of a concept they can relate to. Non-examples benefit readers by presenting instances of what is *not* part of the concept.

When to Use Examples and Non-Examples. Use examples whenever you expect confusion. Confusion might result from a lack of clarity. For instance, you might include an example when describing an abstract concept. Confusion might also result from similarity with other concepts. In these instances, you use both examples and non-examples to help readers determine which examples are instances of the concept and which ones are not. Consider the similarities between ballpoint and roller pens. They seem to be similar to the average person but use a different type of ink and a different type of point on the pen. To the average person this distinction is irrelevant, but to a pen supplier it is significant.

How to Select Examples and Non-Examples. According to Becker, you should carefully select examples to show the range of the concept. That is, you should select examples that show "ends" of the definition as well as its middle. For instance, when explaining the concept dog, you might use Siberian huskies, terriers, and chihuahuas as examples.

You should carefully select non-examples, too. Each non-example should be as closely related as possible to the example it might be confused with. For each non-example, explain why it is not an example. For instance, frozen yogurt is an ideal non-example for ice cream because it looks and tastes like ice cream but is made by a different process.

One technique for presenting examples and non-examples is to present them together as follows:

1. Define the concept.
2. Present a non-example that will not be confused with the concept.
3. Present a non-example that might be confused with the concept.
4. Present an example that is closely related to the non-example presented in step 3. The two should differ only in a main feature.
5. Present an example that significantly differs from the example presented in step 4.
6. Present an example that significantly differs from the example presented in step 5.

Example of What You Should Do

1. Define the concept to be taught.

> A database is a collection of related information. (Show a picture of airline information grouped together.)

2. Present a non-example that will not be confused with the concept.

> An office is not a database. (show picture of an office) (will not be confused with the concept)

3. Present a non-example that might be confused with the concept.

> A recipe is not a database. (show a recipe) (might be confused with the concept)

4. Present an example of the concept that is similar to the non-example just presented.

> A cookbook is a database of recipes. (show a collection of recipes)

Notice how the non-example in step 3 and the example in step 4 are similar. By changing the non-example from one recipe to a collection of recipes, the non-example became an example.

5. Present additional examples of the concept that are very different from the first example.

> A list of all airline passengers is a database. A class list is a database.

6. Present a non-example that is similar to the example just presented. Ask the reader, "Is this an example?"

> Is your score on the marketing training exam an example of a database?
> Readers should answer "no."

Notice how the second example in step 5 and the example in step 6 are similar. By changing the example from a class list to a class score, an example becomes a non-example.

You can continue this process by presenting five more instances, some examples, some non-examples. After presenting each, ask, "Is this an example?" If readers answer incorrectly, explain why the problem is or is not an example of the concept being taught. This is an especially useful technique in tutorial material.

> Is an employee list a database?
> Correct answer is "yes." If a reader says "no," reply that "A database is a collection of related items. All of the items on the list are employees; they're related and collected and therefore an example."

End the presentation with a sample quiz question, in some form other than yes/no.

> From the following list, identify those items that are a database.
> - ☐ Items in the Holiday Catalog
> - ☐ Home address
> - ☐ Telephone numbers for literary society

Procedures

What a Procedure Is. A *procedure* is a set of instructions for performing a task. The *task* can be physical, such as installing a modem, or mental, such as calculating the profit margin on a product. Procedures are among the best-known formats in technical communication.

How Procedures Benefit Readers. Procedures tell readers "how." Specifically, procedures describe the following:

- How to do something, such as change the oil filter in a car.
- How various parts of a task relate to one another, such as how a bill passes through the legislature.
- How something was done. Readers can then verify what happened. For example, a researcher can review the procedures another researcher used to determine whether the analysis is valid.
- How something will be done. Readers can then make necessary plans. For example, a procedure might state the plan for a documentation project. Readers can then determine when they must distribute review copies (or receive them for review) and plan their time accordingly.

When to Write Procedures. Use procedures to present any information that follows a chronological sequence. You can therefore use procedures for information other than those readers actually perform. For example, you can use a procedure to describe how someone completed a task.

How to Write Procedures. A procedure consists of a sequence of actions. The action can be performed physically or mentally (Dick & Carey, 1990). When performed together, these actions complete a larger task.

Do the following to write procedures:

Format — Use numbered lists. That is, begin each step with a number. The number tells readers the sequence for performing the steps. For example:

1. Open the car door.
2. Insert the ignition key.
3. Turn the key to start the engine.
4. Begin to drive.

Length — Because readers can only handle a limited amount of information at any given time, limit procedures to seven to ten steps. If your procedure has more steps, combine several of the steps and present them as a "mini-procedure" within the larger procedure. Present mini-procedures in one of these ways:

- Create a separate procedure with its own title. For example, suppose you are writing a procedure for solving problems with a computer. Few computer problems can be solved in ten steps. So, you might present an initial procedure that helps readers identify the nature of the problem (such as a hardware or software problem), then have separate procedures for solving the different types of problems.

Create a "procedure within a procedure." That is, you might write the mini-procedure as a series of steps within the larger procedure. When writing it, you indent the mini-procedure within the related step of the main procedure. To distinguish the steps of the mini-procedure from those of the main procedure, begin each step of the mini-procedure with a letter. For example:

1. Open the car door.
2. Start the engine.
   a. Insert the key into the ignition switch on the right side of the steering column.
   b. Turn the key away from you.
   c. Listen for the engine to start.
3. Begin to drive.

Content

Consider the following when writing the steps in a procedure:

- Limit each step in a procedure to one action. If the step has several actions, break the step into several smaller steps or write a mini-procedure. Real-world experience indicates that readers stop reading as soon as they encounter the action to be performed in a step.
- Present special conditions for performing the action as a clause at the beginning of a step; do not place these conditions after the action. For example, if readers need to use both hands to hold a piece in place on an assembly line, say it at the beginning of the step, like this:

  6. Holding the piece with both hands, place the assembly in the drying machine.

  Do not write the clause like this:

  6. Place the assembly in the drying machine and hold it with both hands.

  As just mentioned, readers usually stop reading as soon as they encounter the action and perform it. Placing the special conditions before the action is one way to make sure readers see it.

  If readers need knowledge, supplies, or assistance to perform the procedure, state that information in a "Read Me First" section, which precedes the procedure.
- Explain the goal of the procedure as succinctly as possible. Often, a good heading will suffice. For example:

  Procedure for Starting Your Car
- Explain only "must-know" terminology and concepts in a procedure. Readers use procedures to learn how to do

something, not how to understand the concepts and terminology underlying them.

For example, readers must know what a command is before they can enter it. But they do not need to know how a computer processes the command. A procedure for users, then, might define the term "command" but would not explain the steps the system follows to process a command.

- Present supplies needed as a checkoff list. Each checkbox on the list signals readers that they need to do something. For example:

  Before starting the procedure, get these supplies:

  □ Tape
  □ Scissors
  □ Construction paper

- Mention assistance needed in terms of "Who should perform the procedure." For example:

  Who should perform the procedure?
  Technical writers
  Technical editors
  Technical illustrators

  When several people perform a procedure together, state who performs each step and how responsibility passes among people performing the procedure. For example:

  1. The technical writer writes the document.
  2. The technical writer electronically transmits the document to the editor.
  3. The technical editor:
     a. Prints the document.
     b. Identifies the passages that need to be illustrated.
     c. Completes an Art Request Form for each illustration needed.
     d. Submits the Art Request Forms to the technical illustrator.

Tone

Write each item on the list as a self-contained action. For actions that readers are expected to perform, write in the imperative mood. That is, begin with an action verb and tell readers what to do. Research shows that users read these types of directions faster than ones that do not begin with actions (Mirel, 1991).

Turn the power on

or

Subtract expenses from revenues.

For actions that readers are not expected to perform, include the agent for each action so readers know who or what is responsible for that action. (That's another way of saying, write in the active voice.) For example:

1. The Information Development department will distribute the information plan by January 22.
2. The Programming, Product Test, and Marketing departments must submit their review comments on the plan by January 29.

Rather than:

1. The information plan will be distributed by January 22.
2. Review comments must be submitted by January 29.

In this second version, readers do not know who is responsible for performing the actions.

End — End a procedure by telling readers that the procedure is complete. You might write "End of procedure," or be more descriptive, writing something like "Your VCR should now be installed and ready for use."

Examples of What You Should Do. Here's a procedure that readers are expected to perform:

> To choose a task from the action bar without using a mouse:
>
> 1. Press F10.
> 2. Move the cursor to the task you would like to perform.
> 3. Press Enter.

Notice that each step begins with a number and is written as a command to readers.

Here's a procedure that readers are not expected to perform:

> 1. Using the analysis of covariance test, Miller analyzed the data.
> 2. Miller presented the results to Richards and Defoore for verification.
> 3. Richards and Defoore certified the results.

Notice that each step begins with a number and that the agent for each step is identified (the agent is the person who performed the step). Notice, too, that the conditional clause in step 1, "Using the analysis of covariance test," appears at the beginning.

Examples of What You Should Not Do. Notice that the following procedure is presented as a paragraph and that the paragraph is written in passive voice. This tells readers what should be done, not what they should do.

> To choose a task from the action bar without using a mouse: the F10 should be pressed, then the cursor should be moved to the task to be performed. Once the cursor is there, the Enter key should be pressed.

Here is another example of what you should not do. Notice the conditional clause that follows the action in step 1 ("using the analysis of covariance test"). Notice, too, that steps 1 and 2 are written in passive voice; readers do not know who performed those steps.

> 1. The data were analyzed using the analysis of covariance test.
> 2. The results were presented to Richards and Defoore for verification.
> 3. Richards and Defoore certified the results.

Formats for Persuading

What Persuasion Is. *Persuasion* is a means of convincing readers to take a particular action (such as buying a product) or point of view (such as supporting a particular theory).

When Technical Communicators Need to Persuade. Although technical communicators seemingly write about "factual" information (information that readers accept as fact), we often must persuade readers to a point of view.

For example, when describing a procedure or a new concept, we must persuade readers that the procedure or new concept has value to them. As mentioned in Chapter 1, readers must see the value of information before acting on it.

Some documents are intended solely for persuasion. If readers act as you want them to after reading your document, then you have successfully persuaded them. Two types of documents are intended primarily for persuasion:

- Sales information. After reading sales information, your audience should be persuaded to purchase a product or service.
- Scholarly and professional articles. After reading these articles, your audience should be persuaded to support a particular theory or point of view.

The following sections explain how to present these types of information.

Sales Information

What Sales Information Is. *Sales information* provides readers with information on products and services. The information might be intended to persuade readers to make the purchase immediately or to continue their consideration of the product or service.

How Sales Information Benefits Readers. Sales information helps readers decide whether to purchase particular products and services.

When to Use Sales Information. Use sales information in sales brochures, general information manuals, mall kiosks, videotapes and other information that is intended to sell your products or services.

Most often, sales information is directed at paying customers and prospective customers outside your organization. For example, you might publish a flyer describing the desktop publishing software your company sells. Through this flyer, you hope to generate cash sales for your organization.

Sales information can also be published for people working within the same organization. Usually, such "internal" sales information is intended to make others aware of a little-known department or service. For example, you might publish a brochure that tells other departments in your organization about the services you provide. Through this brochure, you hope to increase the number of projects your department handles.

How to Write Sales Information. Sales information is a lot like pesticides: as insects develop immunities to pesticides, so people develop an immunity to sales information (Dillon, Madden, & Firtle, 1987, p. 41). Although sales information is essential for communicating information about your organization to prospective customers, the information must be presented in realistic terms and must be sensitive to customers' needs.

Sales information has these parts:

1. Grabber, which was described earlier.
2. Value section, which explains the benefit of the product or service to the customer. For example, the value section of a large sales proposal might explain how a product meets all the requirements the customer has established. The value section for a sales brochure on a new VCR might explain that it is easier to program than other VCRs. The value section for a sales video on a new X-ray machine might show how the machine takes more complete and accurate X-rays than previous models.

3. Features section, an optional section that provides detailed information on the features of the product. People often confuse features with value. A feature is a characteristic of the product, such as a pre-programming feature of a VCR. A value explains why the feature is useful to customers, such as letting customers record shows while they are out of their homes. The features section follows the value section because customers need to know what value the product or service holds for them before they care about specific features.
4. Call to action, which tells readers what to do next and how to do it, such as providing a telephone number to call for more information or a list of stores where customers can purchase the product or service.

   The call to action provides you with a means of measuring the effectiveness of your sales information. If you tell people to call or to order and to mention this piece of information when they do so, you can determine how many people actually respond to the call.

As mentioned earlier, the primary purpose of sales information is persuading readers to purchase your product or service. That means you should generate interest in and enthusiasm for your product. That's challenging to do because readers have a limited attention span for sales information. Some ways to overcome this challenge are:

- Presenting information clearly and succinctly. Present only the most important information—the information needed to choose your product or service. You might offer a deferred billing service, but if customers don't need that information to purchase the product, you should not include it in the sales brochure.
- Set realistic expectations for your product or service. Unrealistic expectations usually destroy consumer confidence in your product or service because they lack credibility or because they make promises that are not fulfilled. Therefore, pay careful attention to language when writing sales information. Specifically:
  - Avoid superlatives, such as "best," "tops," "number one," and the like. If you must use these terms, provide an outside verification for your claim and cite the source. For example, "according to a J.D. Powers survey of new car drivers, Brand X is tops in customer satisfaction."
  - Avoid adjectives. Use concrete nouns and verbs instead. For example, don't say "the powerful computer;" say "a computer that can handle 300 users at once."

In other words, stick to the facts that can be verified.

Example of What You Should Do

Grabber

Come to the Communications Crossroads: the 39th Society for Technical Communication Annual Conference in Atlanta, May 10–13, 1992.

Value Section

Learn ways to become more effective at your job. Learn:

- Ways to effectively manage your projects
- Research-based techniques for effectively communicating information
- Technologies for reducing publishing costs and time

And do all this at a cost that can't be beat. The $350 registration fee* for the three days of the STC Annual Conference is less than similar international conferences for similar audiences.

Features Section

The STC Annual Conference fee provides:

- Two general sessions featuring well-known writers
- Over 170 sessions on writing and editing, visual communication, research and technology, management and education, and professional development. You can attend as many as 11 of these sessions.
- Entrance to the Conference Expo, where you can see the latest products and services available to technical communicators.
- Breakfast each day.
- A reception the first evening.
- A conference *Proceedings* (not provided with reduced-fee registrations).

Call to Action

Complete the attached registration form and send it along with your check by May 1 to:

Annual Conference Registration
Society for Technical Communication
901 North Stuart Street
Arlington, VA 22203

*Reduced fees for one-day attendance, students, and seniors

Note the following about this example:

- The grabber uses a catchy phrase to interest readers and incorporates a "lead" format. The catchy phrase is the theme for the conference.
- The value section focuses on two characteristics of the conference:
  - How information learned at the conference can benefit readers on the job
  - How the conference is competitively priced with similar conferences
- The call to action asks readers to register for the conference and tells them how to do so.

Example of What You Should Not Do

Grabber

The best diet in the business!

Value Section

This easy-to-follow diet consists of:

- Pre-packaged dinners
- Pre-packaged snacks
- Weekly diet plan
- Unlimited visits to the office

Features Section

Lose as much as 10 pounds in your first week!
This is a doctor-approved diet.

Call to Action

Try our diet today at the location nearest you.

Note the following about this example:

- The grabber might get your attention but it makes a bold claim that should be supported with evidence. Readers will want to know who says this is the best diet in the business.
- The values and features are transposed. The value to readers is that they can lose as much as 10 pounds in one week. The features are the pre-packaged meals, the weekly plan, and the office visits.
- The call to action tells readers what to do but doesn't tell them how. The call to action merely says "the location nearest you"; it doesn't tell readers where those locations are.

Scholarly and Professional Articles

What Scholarly and Professional Articles Are. Scholarly and professional articles describe topics that support a particular theoretical viewpoint. These articles are intended to be read by other professionals in the field. Although the stated purpose of these articles might be to inform readers about a theory or research study, their actual purpose is to persuade readers that a certain belief—or hypothesis—is based on evidence uncovered in research or that readers should follow a certain theory, based on evidence presented in the paper (Boiarsky, 1991, p. 96).

How Scholarly and Professional Articles Benefit Readers. These articles benefit readers in many ways:

- They explain current work methods (and why these work methods are superior). For example, articles for technical communicators explain techniques for writing hypertext, using color, and conducting author-editor conferences.
- They explain theories that guide work in the field. These theories provide a framework for approaching work tasks and for research. For example, the "naturalistic" theory proposes that everything in the world must be understood in its context (Lincoln & Guba, 1989). As a result, research should be conducted in the field rather than in a laboratory. Chapter 11 describes field testing.

When to Write Scholarly and Professional Articles. Write a scholarly or professional article whenever you have something significant to share with your colleagues:

- An idea
- A method or technique you have researched or perfected
- Results of research that you have conducted

How to Write Scholarly and Professional Articles. Scholarly and professional articles are intended to persuade readers to a particular point of view or present information readers can use on their jobs.

Whatever their purpose, scholarly and professional articles follow a general framework:

1. Informative abstract, which explains the main points of the report and is similar to a summary.
2. Introduction, which was described earlier in this chapter.
3. Theoretical framework (literature review), which explains the theory underlying the work method or research described in the article. In many instances, it is called a literature review because the

theory underlying the article is based on, or is a response to, information already published.

4. Statement of position (called a thesis statement in an essay and a hypothesis in a report of an experiment), which is a one- or two-sentence declaration of what you are trying to prove in the rest of the article. For example, in a report analyzing a marketing brochure, the thesis might be:

   The creative approach and unique call to action have made this marketing brochure an effective one.

   The thesis is that the brochure is effective for two primary reasons. Support for these two reasons becomes two of the main sections in the report. A third main section might present other proof that the brochure is effective.

   The hypothesis of a research report might be:

   People using the information in manual A will be able to perform the same procedure in 30% less time than people using the information in manual B.

   The research data that you present should support this hypothesis. For example, you might present the time each reader needed to perform the procedure and the manual each reader used.
5. Presentation of information, which presents the work method described or the research conducted.
6. Application of the information, which presents examples of the work method or an analysis of the research results.
7. Summary or conclusion, which recaps the main points of the report or provides a concluding thought (or both).

Scholarly and professional articles persuade readers through evidence that supports the position stated in the article. For example, a professional article might claim to present ten techniques for writing audiovisual scripts. The author must then present ten tips that help with audiovisual scriptwriting. Similarly, a scholarly article might claim that one writing technique is more effective than another. The evidence might include the results of tests in which people performed tasks written in these two techniques. Evidence includes:

- Real-world experience
- Facts learned in a review of information on the subject
- Results of research conducted in the subject area
- Opinions from experts

The evidence must be strong enough to withstand the scrutiny of experts. That's why claims stated in most academic articles are supported by a footnote or citation, and quotes in professional articles are attributed to a source.

Formats for Referencing

What Referencing Is. Referencing is a technique readers use when looking for a specific piece of information.

When Technical Communicators Need to Use Referencing. Use referencing when readers need specific pieces of information, such as which research technique to use when performing a genetic engineering test or the command used to make a backup copy of a diskette.

Formats available for referencing include:

- Reference entries, which provide information about a part of a system
- Glossaries, types of reference entries for presenting information about terms

Reference Entries

What a Reference Entry Is. A *reference entry* provides information about a part of a larger system. Reference entries are usually written for readers who have experience with the subject matter but need a refresher on the specifics. They are contained in a reference passage.

The subject should be specific enough that you can cover it in a limited space. For example, commands in the BASIC programming language make ideal reference entries. These are brief and to the point. An entry on the causes of the Depression is not appropriate. This entry would cover a lengthy subject requiring extensive analysis.

How Reference Entries Benefit Readers. Reference entries benefit readers by providing them with a single source of information about a specific topic. Related topics are located elsewhere in the document to keep each reference entry as brief as possible.

Reference entries are usually listed in alphabetical or numerical order so readers can easily find them. Readers might read individual entries in their entirety but rarely read all the entries in a reference document.

When to Use Reference Entries. Use reference entries to describe subjects readers may be familiar with but have not memorized (and probably won't). Specifically, consider using reference entries in the following instances:

Type of Information	Why You Should Use a Reference Entry
Commands for programming languages and other types of software	Readers usually need information about a particular command; a reference entry helps readers easily find it.
Definitions	Readers usually want a definition of an individual term. A reference with definitions is called a glossary or dictionary. Glossaries are described later in this chapter.
Policies	Readers usually need information about a particular policy; they prefer not to read about all of them at one time. A reference source of policies is usually called a policy manual.
Style	Readers need information about a particular rule of style, such as how to punctuate bibliographic entries. A reference source on style is called a style guide.

How to Write Reference Entries. Reference entries should have these characteristics:

- Provide all the information readers need about the subject
- Provide similar information in each entry for consistency

When planning a group of reference entries, you first determine which information readers need. Then, you develop a consistent format to use for each entry.

The actual information you provide in an entry varies, depending on the information you are presenting and the readers you are serving. Common reference entries include the following:

Type of Reference Entry	Information Provided in the Entry
Commands for programming languages and other types of software	1. Command name. 2. Command syntax (order in which readers should enter the command and its parameters) 3. Definition of the command 4. Explanation of parameters (required and optional values entered with the command) 5. Considerations for using the command (such as combinations of parameters you cannot use). 6. Examples of the command being entered (with various combinations of parameters) 7. Examples of output produced by the command (if any)

(*continued*)

Type of Reference Entry	Information Provided in the Entry
Definitions	See "Glossaries" later in this chapter for instructions on how to write this special type of reference entry.
Policies	1. Name of the policy. 2. Explanation of the policy. 3. Instructions for administering the policy.
Style	1. Rule. 2. Explanation of the rule including a rationale. 3. Example.

Readers expect consistency among reference entries. When you write these entries, therefore, make sure that each entry in a document has all the same parts. Similarly, within reference entries, you should include only the information your readers need. For example, if you are writing a policy manual for employees who do not have the authority to administer the policies, you omit a section on administration from each reference entry. This information would be included only when writing a manual for administrators.

Finally, note that readers should already be familiar with technical terms discussed in the reference entry. Provide a glossary if you do not expect readers to be familiar with technical terms.

Examples of What You Should Do. A command description:

RUN Command

```
RUN \directory\program-name
```

Starts the program whose name is entered with the command.

\directory is the name of the directory in which the program is stored. If you do not enter a directory name, system assumes the program is stored in the root directory.

program-name is the name of the program to be run. You must enter a program name.

Considerations

None.

Examples

```
run \mydir\account
```

Runs the program named account in the directory named mydir

```
run wordproc
```

Runs the program named wordproc.

A policy:

Educational Leave

About the Policy: Allows employees to take up to 2 years off without pay to complete a degree program at an accredited university. Employees on educational leave are guaranteed a comparable job upon return if they receive passing grades in all of their courses.

Administering the Policy:

— Eligibility: Employees who have a minimum 5 years with the company, who have been accepted into a degree program at an accredited university, and who have satisfactory work performance.

— Benefits and Salary: Employees continue to receive health and dental benefits when on leave. They do not receive salary or other benefits.

— Part-Time Work: Employees on an educational leave may work up to 20 hours weekly. In this capacity, they will be considered supplemental employees. See "Supplemental Employees" for information about this program.

Glossaries

What a Glossary Is. A *glossary* is a section in a document that contains definitions of terms used in unfamiliar ways.

How a Glossary Benefits Readers. Glossaries provide readers with a place to find terms they might not be familiar with. Readers especially appreciate glossaries when they encounter an unfamiliar word (perhaps a word they read once) and cannot easily find a definition in the text. Readers also appreciate glossaries when they encounter a word they think they should know but can't remember its definition.

When to Include a Glossary. Include glossaries in any document that uses a lot of technical terms or terms that have unfamiliar definitions for some readers.

How to Write a Glossary. Writing a glossary involves the following:

1. Select the terms to include. At the least, you should include all new terms introduced in the document. At the most, you should include all terms that readers might encounter but whose definition might not be remembered.

2. Write the entries. A glossary entry includes:
   a. The term. Highlight the term, preferably in boldface type.
   b. The definition. When possible, use the same definition used elsewhere in the document. This definition should clearly distinguish the term from similar ones.

   Optionally, a glossary entry includes:

- The part of speech, which tells readers whether, for example, the term is a noun or a verb. This is especially useful information for terms that are familiar to readers but are used in unfamiliar ways. Place the part of speech immediately after the term.
- Examples, which clarify abstract terms and terms that are similar to others.
- Synonyms, which are alternative terms with the same meaning as the defined term.
- Related terms that readers should also be familiar with. Begin the list of related terms as "see also." Make sure all of the "see also" terms are in the glossary.

Note that glossaries differ from dictionaries. Glossaries present only the uses of those terms that are relevant to the document. Dictionaries provide more complete definitions, presenting all uses of a word.

Examples of What You Should Do. In the glossary entry that follows, note that the definition is highlighted and that the first letter of the term is capitalized. You do not need to capitalize the first letters of terms, but you should be consistent in format. That is, if you capitalize the first letter of one term, you should capitalize the first letter of them all.

> **Copy:** a command used to copy files. Use this command to copy a single file or a group of files.

Note the synonyms included in the following example:

> **Computer-based training:** Training that students receive through a computer. Also called computer-aided instruction, computer-assisted instruction, and computer-based instruction.

Note the related terms in the next example. The list of related terms begins with the words "See also."

Introduction: A paragraph or group of paragraphs at the beginning of a passage that grabs and holds readers' attention, presents an overview of the upcoming passage, and relates the passage to readers' needs. Also see *formal introduction, grabber,* and *lead.*

Writing Summaries

What a Summary Is. A *summary* reinforces the main points of a section or document by giving readers enough detail to remember the information. This redundancy helps readers (Pieper, 1991).

Summaries can appear at the beginning or the end of a document:

- *Executive summary* appears at the beginning of a document
- *Concluding summary* appears at the end of a section or of a document

How Summaries Benefit Readers. Summaries benefit readers in these ways:

Type of Summary	How It Benefits Readers
Executive summary	Presents the main points of the document so readers can get a quick overview of the key points covered and determine whether they need to read the document in its entirety. Executive summaries are especially useful in reports.
Concluding summary	Gives readers one last reminder of what they should recall after reading the section.

When to Include Summaries. Include summaries when readers need an overview or recap of the information, such as a guidebook, a report, or a video. Do not include summaries for information that readers will only use as resource, such as a reference manual or hypertext references.

Type of Information	Type of Summary to Use
Reports	Executive summary
Instructional (including course materials, guidebooks, textbooks, and videos)	Concluding summary

How to Write Summaries. You write executive and concluding summaries in a similar way. These types of summaries generally cover some of the same material as an introduction, but provide more specific information about the contents.

For example, if the introduction tells readers that the chapter will present three ways of doing something, the concluding summary names those three methods. The summary might name the steps in the procedure. The summary might repeat definitions of significant terms. Note that each chapter in this book ends with a summary.

Similarly, an executive summary should describe the problem addressed in the report and its implications, identify the solutions available, state the preferred solution, and why it is recommended (Vaughn, 1991).

Examples of What You Should Do. Following is an example of a concluding summary:

Chapter Three: Using the Word Processor

This chapter explained how to:

- Enter text by typing on the text Entry Panel.
- Copy, move, and delete text by using the Edit window and specifying the correct option.
- Set margins and tabs by using the Format window.
- Emphasize text using underlining and bold and italic type by using control key combinations.
- Save text in a file by using the File window and choosing a file name and options.

Notice how this summary lists the main points of the chapter and provides the primary details about each. For example, the description of saving text not only identifies the task but also the key steps in performing it.

Following is an example of an executive summary:

The company needs increased efficiency and economy in performing routine drafting and conceptual designs. This study shows that our engineers are 50% less productive than those working at Barnes and Wells Engineering. To resolve this problem, the drafting department should purchase a new computer-aided drafting (CAD) system.

Example of What You Should Not Do. Although the following summary repeats the main point of the chapter, it fails to provide enough detail for readers to remember how to use the main functions.

Chapter Three: Using the Word Processor

This chapter explained how to use the main functions of the word processor.

WRITING GENERICALLY

What Writing Generically Means. Much technical information is revised and published again. For example, most computer software is regularly updated, requiring the manuals that describe the software to also be updated. In addition, Levine, Pesante, and Dunkle (1991) compare written documents to software in that entire parts of documents—from phrases to entire chapters—can be reused for new purposes either in part or in whole.

By considering that information might be reused, you can simplify the process and cost of revising and reusing the information when you first write it by writing generically (Holden, 1986). This means that you avoid using unnecessary details whenever possible, such as names, numbers, and cross-references.

Consider this example. Instead of writing "the following *three* items," you might write "the following items" so if the list of items expands to four when the document is revised, the lead-in to the list does not need to be revised.

In most instances, writing generically does not hinder the reading process. Writing generically might even help readers by keeping them focused on the key details rather than trivial ones. If, however, writing generically removes essential information, don't write generically. For example, if readers are forgetting to check all five items when performing a safety check and do not do so unless they are told how many items to check, make sure you include the number five.

When You Should Write Generically. Write generically whenever a document might be revised or reused. Usually you are told that a document might be revised. Even when you're not told, you can make an educated guess. For example, information about software is almost always revised. Sales information is often revised with the introduction of new or changed products.

Benefits of Writing Generically. Writing generically reduces:

- The amount of information that needs to be revised because passages that contain no technical information do not need revision
- The likelihood of errors resulting from failing to revise non-technical phrases that required revision
- The cost of revision and reuse because fewer pages are affected by changes

Writing generically, then, is writing with an eye toward the future. By avoiding unnecessary detail in an early draft, you can avoid burdensome revisions in future drafts.

How to Write Generically. Writing generically involves questioning whether details are necessary. Here are common examples of unnecessary detail:

- Identifying the number of items on a list. In a revision, the number often changes. For example, instead of saying:

  The system has three elements

  Say:

  The system has these elements

  At the next revision, the system might be simplified to contain only two elements.
- Stating model numbers when a generic product name will do. In a revision, the model numbers often change. For example, instead of saying:

  If you are using all of the models–models 1, 2, and, 3

  Say:

  Whatever model you use

  At the next revision, model 2 might be discontinued and a model 4 might be added.
- Referring to other parts of a document by number. Instead, refer to them by name. For example, rather than saying:

  Chapter 6

  Say:

  Installing the System

  In a future edition, this chapter might become chapter 2.

Examples of writing generically

- "This dishwasher offers many benefits." If a later version offers benefits not available when the brochure is written, the sentence would not need to be revised.
- "In the Conclusion" If the section number for the Conclusions changes, the sentence does not need to be revised.

Examples of when not to write generically

- "Sign all five parts of the insurance agreement or the agreement is not considered valid." Five is a significant detail in this sentence. If readers do not sign all five parts of the agreement, they face serious consequences.
- "Autostart is available only on Model B33." The model number is a significant detail because autostart is available only on that model.

OTHER ISSUES IN PRESENTING INFORMATION

Using an appropriate format to present information is key to readers' understanding it. As you write the information, however, you should also be aware of these concerns:

- You must maintain technical accuracy. People trust what they read and, as a writer, you must uphold that trust.

  To ensure technical accuracy, make sure that information is reviewed by knowledgeable experts. In some instances, you can serve that role. In other instances, you need to ask technical experts to read the information and make sure it is correct. For large documents, you should formally schedule reviews with technical experts to make sure that they take this reviewing task seriously and to make sure that they schedule sufficient time to review your work. The more comments you receive, the more likely your draft was fully reviewed. Chapter 3 suggests strategies to follow when scheduling reviews.
- Present only the information that people need—no more, no less. As mentioned in the introduction to this book, this is the information age and readers are inundated with information. One of your responsibilities as a technical communicator is ferreting out the necessary information from the unnecessary. When doing so, consider your readers' needs. If readers need the information, include it. If readers don't need the information, drop it.

  As you go through the writing and review process, you might be asked to include information that does not meet readers' needs. Do not include it. Be careful, though, when determining which information readers need. On the one hand, few readers actually read a warranty, but they do need it. On the other hand, few computer users care about the hardware registers in their computer—or have a need to know it.

EXAMPLES OF CHOOSING FORMATS FOR PRESENTING INFORMATION

The past several sections have described various formats for presenting information. A single document, however, uses several formats. To show you how you select formats for different parts of a document, consider this extended example.

Imagine that you have been asked to develop an online travel guide. To organize the information, you determine that the guide should tell readers:

- The rules and regulations for company travel
- The hotels they are authorized to use

- The method for making airline and hotel reservations through the online reservation system

Now, you need to determine how to present the information in each part of this online guide. Section by section, consider what you might do.

Section	An Appropriate Presentation Format
Rules and regulations	This is reference material, so you might create a reference section. This section might include several types of entries, such as rules for airline travel, for hotel stays, for meals, for allowable living expenses, and for filing reimbursements.
Authorized hotels	This is reference material, too, but not the same type of information as the rules and regulations. You should therefore create a different type of entry. Talk with your readers to determine how they would like to have this information presented. For example, if readers are going to consult this information to determine where to stay when visiting a particular city, you might list hotels by city. The entry for each city might provide the names, addresses, telephone numbers, nightly rate, and proximity to the local office for each hotel listed.
Instructions for making reservations	This is a procedure and should be written as such. Because the information is presented online, you might present the instructions in two ways to accommodate the needs of two types of users: 1. A series of prompts, letting novices proceed one step at a time. Readers must complete each step before advancing to the next. 2. A list of instructions, providing experienced readers with a quick reference. Users do not interact with this list.

SUMMARY

This chapter began with a discussion of craftsmanship in technical communication, which refers to the ability to create precise and accessible documents. It is also the ability to choose the most appropriate format for presenting information in the least amount of space.

When choosing the most appropriate means for presenting information, choose a format best suited to your purpose. This chapter described several of the most commonly used formats for presenting information verbally:

- Formats for introducing passages, including:
  - Prefaces, which explain the purpose of a document.
  - Openings, which tell readers the value of the upcoming information and present an overview of it.

- Grabbers, which are catchy phrases intended to grab and hold readers' attention.
- Leads, which are opening paragraphs that present the main points of a presentation: the five Ws and H and, in memos, a call to action.

- Formats for presenting specific types of material:
  - Instructional material, including:
    - Concepts which, according to Becker, "can be defined by ... properties ... common to the concept instances and not common to non-instances."
    - Analogies, which are comparisons of unfamiliar concepts to familiar ones that explain what the similarity is.
    - Examples, which provide real evidence of a concept.
    - Procedures, which are explanations of tasks.
  - Persuasive material, including:
    - Sales information, which provides readers with information on products and services. The information might persuade readers to make the purchase or to further consider the product or service.
    - Scholarly and professional articles, which describe work methods, recent research, and theories.
  - Referencing, including:
    - Reference entries, which describe in detail parts of a system.
    - Glossaries, which clearly define terms.
- Formats for summarizing by presenting or repeating the main points of the passage.
- Techniques for writing generically to avoid unnecessary detail so that revisions can be made at a reduced cost and effort.

This chapter concluded with a discussion of accuracy and of controlling the amount of information readers encounter: two key issues in presenting information verbally.

LIST OF TERMS

analogy A comparison of an unfamiliar concept to a familiar one that explains what the similarity is.

call to action Tells readers what is expected of them after reading the memo.

concept According to Becker, a concept "can be defined by . . . properties . . . common to the concept instances and not common to non-instances."

example An instance of a concept.

explanation A sentence, paragraph, or groups of paragraphs that defines a concept and presents associated rules and definitions.

genre A format for presenting information. A genre occurs when different communicators address similar situations in a similar way. Specifically, communicators writing in a particular genre include similar types of content and write in a similar style.

glossary A section in a document that contains definitions of terms used in a document.

grabber A sentence, paragraph, or group of paragraphs at the beginning of a passage that grabs and holds readers' attention.

introduction A paragraph or group of paragraphs at the beginning of a passage. Also see *grabber, lead,* and *opening*.

lead A paragraph at the beginning of a passage that presents all the main points of the passage.

non-example In instruction, something that is not an example of the concept being taught but might be confused with the example.

opening A paragraph or group of paragraphs at the beginning of a passage that provides an overview of upcoming information and relates that information to readers' needs.

persuasion Means of convincing readers to take a particular action or point of view.

preface A section at the beginning of a document that explains what the document is about, who its intended readers are, what readers should already know before reading the document (if appropriate), how readers should use the document, and other useful information about the document.

procedure A set of instructions for performing a task. The task can be physical, such as installing a modem, or the task can be mental, such as the procedure for calculating the profit margin on a product.

proposal A special type of sales information intended to acquire new business and usually prepared in response to a request from a prospective customer. Also see *sales information*.

reference entry Provides detailed information about one part of a system.

sales information Tells readers about products and services. The information might persuade readers to make the purchase or to further consider purchasing the product or service.

summary Means of reinforcing the main points by giving readers enough details to remember the information.

task An action that can be performed physically or mentally.

REFERENCES

Barakat, R. (1991). Developing winning proposal strategies. *IEEE Transactions on Professional Communication, 34,* 130–139.

Becker, W. (1977). Teaching reading and language to the disadvantaged—what we have learned from field research. *Harvard Educational Review, 47,* 518–543.

Becker, W. (1986). Applied psychology for teachers. Chicago: Science Research Associates.

Boiarsky, C. (1991). Relating purpose and genre through James Britton's functional categories. *The Technical Writing Teacher, 18,* 95–103.

Braun, M. (1991). Editorial: Personalizing the information revolution. *T.H.E. Journal, 19,* s.s.-1.

Brusau, C., Alred., G., & Oliu, W. (1976). *Handbook of technical writing.* New York: St. Martin's.

Dick, W., & Carey, L. (1990). *The systematic design of instruction.* (3rd ed.) New York: HarperCollins.

Dillon, W., Madden, T., & Firtle, N. (1987). *Marketing research in a marketing environment.* Homewood, IL: Irwin.

Flower, L. (1978). *Problem-solving strategies for writing.* Pittsburgh: Carnegie-Mellon University.

Foss, S. (1989). *Rhetorical criticism: Exploration & practice.* Prospect Heights, IL: Waveland.

Frank, D. (1985). *Silicon English: Business writing tools for the computer age.* San Rafael, CA: Royall.

Gagné, R. M. (1985). *The conditions of learning and theory of instruction.* (4th ed.). New York: Holt, Rinehart and Winston.

Holden, N. (1986). Generic and other types of common writing. *Proceedings of the 33rd International Technical Communication Conference.* Detroit: Society for Technical Communication, pp. MPD 145–147.

Horn, R. (1983). Class handout for course, *The information mapping course for writing procedures, policies, and documentation.* Waltham, MA: Information Mapping, Inc.

Horton, W. (1991). *Illustrating computer documentation: The art of presenting information graphically on paper and online.* New York: John Wiley.

Jonassen, D. (1982). Advance organizers in text. In D. Jonassen, (Ed.). *The technology of text: Principles for structuring, designing, and displaying text,* (pp. 253–276). Englewood Cliffs, NJ: Educational Technology Publications.

Levine, L., Pesante, L., & Dunkle, S. (1991). Implementing the writing plan: Heuristics from software development. *The Technical Writing Teacher, 18,* 116–125.

Lincoln, Y., & Guba, E. (1989). *Naturalistic inquiry.* Newbury Park, CA: Sage.

Mirel, B. (1991). Critical review of experimental research on the usability of hard copy documentation. *IEEE Transactions on Professional Communication, 34,* 109–122.

Newby, T., & Stepich, D. (1991). How to write analogies. *Proceedings of the 1991 National Society for Performance and Instruction Conference,* Los Angeles.

Pieper, G. (1991). Effective roadmaps or signs of construction ahead: *The Technical Writing Teacher, 18,* 126–131.

Price, J. (1984). *How to write a computer manual.* Menlo Park, CA: Benjamin/Cummings.

Schroello, D. (1990). How to market training and development programs. Presented at the 1990 Best of America Conference, New York.

Vaughn, D. (1991). Abstracts and summaries: Some clarifying distinctions. *The Technical Writing Teacher, 18,* 132–141.

Weiss, E. (1991). *How to write usable user documentation.* (2nd ed.). Phoenix: Oryx.

SUGGESTED READING

Babbie, E. (1983). *The practice of social research.* (3rd ed.). Belmont, CA: Wadsworth.

Barnum, C. (1986). *Prose and cons: The do's and don'ts of technical and business writing.* Englewood Cliffs, NJ: National (Prentice Hall).

Carliner, Saul. (1987). Lists: The ultimate organizer for engineering writers, in D. Beer, (ed.), *Writing and speaking in the technology professions: A practical guide* (pp. 53–56). New York: IEEE Press.

Carliner, Saul. (1990). Elements of editorial style for computer-delivered information. *IEEE Transactions on Professional Communication, 33,* 38–45.

Horton, W. (1990). *Designing and writing online documentation.* New York: John Wiley.

Olsen, M. (1984). Terminology in the computer industry: Wading through the slough of despond. *Proceedings of the 31st International Technical Communication Conference.* Seattle: Society for Technical Communication, pp. WE 200–203.

Reed, S. (1982). *Cognition: Theory and applications.* Monterey, CA: Brooks/Cole.

Schroello, D. (1991). Seven steps to better internal marketing. *Data Training, 10,* 12–18.

Strunk, W., & White, E. B. (1979). *Elements of style.* (3rd ed.). New York: Macmillan.

Van DeWeghe, R. (1991). What is technical communication: A rhetorical analysis. *Technical Communication, 38,* 295–299.

Wurman, R. (1989). *Information anxiety: What to do when information doesn't tell you what you need to know.* New York: Bantam.

CHAPTER 6

PICTURES PLEASE — PRESENTING INFORMATION VISUALLY

William Horton

William Horton Consulting

WHY USE GRAPHICS?

> "What is the use of a book," thought Alice, "without pictures or conversations?"
>
> Lewis Carroll, *Alice's Adventures in Wonderland*

Until now you have only explored how to communicate with words. But words are not the only means, or even the best means, of communicating ideas. Sometimes visual formats called *graphics* are more effective tools than words. Here's why.

Graphics Communicate What Words Cannot. Some concepts defy words. Other concepts can be expressed in words but are more efficient in pictures. Still others must be translated to visual images before they are understood even though they are expressed in words. For example:

Smoking permitted in the three rows of tables between 2nd and 5th columns.

Graphics Are Understood More Quickly Than Words. Properly designed graphics make their main point in a glance. They do not have to be read, analyzed, and interpreted. With graphics, comparisons become automatic and relationships obvious.

Consider these two ways of representing the mutual likes and dislikes of a group of people. Which form makes answering the following questions quicker?

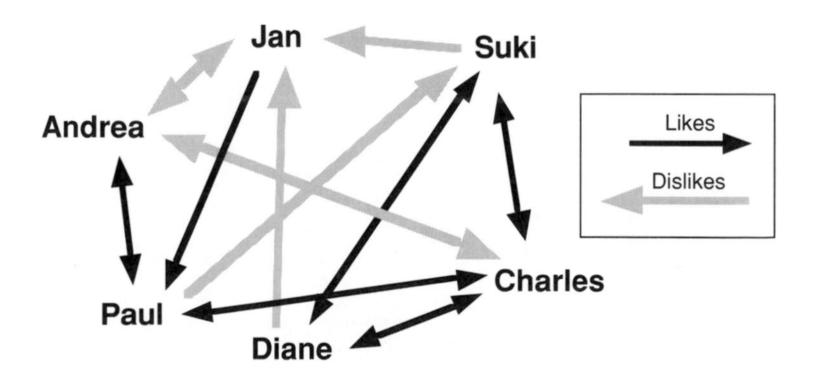

1. Who is the most popular?
2. Who is the least popular?
3. Is there a clique of people who all like each other?
4. Which people have their likes and dislikes reciprocated?

Concrete Images Are Remembered Better Than Abstract Concepts. Do you, like most other people, remember faces better than names? Tests have shown that we have almost unlimited recognition memory for graphic images and that concepts remembered visually are recalled better than those remembered verbally. Try this test:

1. Try memorizing these pairs of words by forming a sentence containing both words.

radio	hand
shoe	river
house	bug
knife	flower
salt	bean
tire	candle
sofa	car

   Now cover the right column and try to recall the words associated with each word in the left column.

2. Next repeat the procedure on the following list of words, but this time try memorizing these word pairs by forming a mental image including both objects.

snake	fireplug
chimney	boat
drum	rabbit
king	garage
fish	wheel
octopus	airplane
cow	flower

3. How well did you recall each list? What does this suggest about how you remember information?

Graphics Record Information Concisely. Even great writers profess the concise power of graphics:

> A picture shows me at a glance what it takes dozens of pages of a book to expound.
>
> Ivan Turgenev

This power is not lost on business leaders.

> The higher one looks in administrative levels of business, the more one finds decisions are based on tabular or graphic formats.
>
> Norbert Enrick

A study by the Wharton School of Management at the University of Pennsylvania found that speakers who used visual aids were twice as successful in achieving their goals as speakers who did not use visual aids. Participants in meetings where visual aids were used retained five times as much information, and these meetings took only half as long.

Pictures Entice and Seduce Readers. Most magazines and paperback books use an attractive graphic on the cover. Even staid technical journals, such as the Journal of the American Medical Association, are using graphics to entice readers to pick up the publication and seduce readers into reading the articles. Similarly, graphics hold attention once people look inside a publication. Consider all the photos, charts, and illustrations in news magazines and newspapers.

Thoughts Are Visual.... In *The Ego and the Id,* Freud noted the visual nature of thought:

> ... it is possible for thought processes to become conscious through a reversion to visual residues.... Thinking in pictures... approximates more closely to unconscious processes than does thinking in words, and is unquestionably older than the latter both ontogenetically and phylogenetically.

This importance was echoed by Albert Einstein who, in a letter quoted in *The Creative Process,* edited by Brewster Ghiselin, described his own thought process:

> The words of the language, as they are written or spoken, do not seem to play any role in my mechanism of thought. The psychical entities which seem to serve as elements in thought are certain signs and more or less clear images which can be "voluntarily" reproduced and combined.... The above mentioned elements are, in my case, of visual and some of muscular type. Conventional words or other signs have to be sought for laboriously only in a secondary stage, when the mentioned associative play is sufficiently established and can be reproduced at will.

... Even in Words. Consider, finally, the importance of vision and seeing throughout our language. We speak of someone as being "far sighted" or "visionary" or a "seer." When we agree we see "eye-to-eye." We are on the "lookout" for bargains and we relish the "sight for sore eyes." Something that gets our attention is "eye catching" or "eye opening." We trust "eyewitnesses" and hire "private eyes." After "looking into" something we develop our own "viewpoint."

In summary, graphics

- Communicate what words cannot
- Are understood more quickly than words
- Are remembered better than words
- Record information concisely
- Entice and seduce readers

Using graphics to express ideas requires as much thought and consideration as using words. This chapter explains how to use graphics effectively as a communication tool. Specifically, this chapter

1. Explains how to select a strategy for communicating with graphics
2. Explains how to implement the graphics strategy by turning words into pictures
3. Helps you develop a repertoire of graphics skills by describing
   a. Eight types of graphics and when you should use them
   b. Five techniques for increasing the quality and effectiveness of graphics

GETTING STARTED: ANALYZING, PLANNING, AND SELECTING A STRATEGY

Communicating with graphics requires as extensive a repertoire of skills as communicating with words. As words can convey varying amounts of information and meaning, so can graphics. As successful prose requires careful thought and a clear plan, so do successful graphics. Before you put pencil to paper, think about who will read the graphic and what type of information it will convey. This section will help you plan successful graphics by explaining how to choose the right amount and type of graphical information, first for your audience and then for the content.

Choose the Right Amount of Graphical Information for Your Audience

Before you actually develop your graphics, you first need to determine how complex your graphics can be. This depends on who your audience is and what type of information you are presenting. First, consider the audience. Readers' jobs, experience, language, and culture all affect how you must tailor graphics for them.

Technical Specialization of the Readers. First, consider the technical specialization of the readers. Is the graphic for the general public, for technicians, or for technical professionals?

If the Graphic Is for	Present This Type of Information	And This Amount of Detail	And Use These Symbols
Professionals, such as scientists, engineers, or programmers	Technical, perhaps showing abstract relationships	Moderate to high (if well designed and clearly organized)	Specialized symbols—but only if explained or well known throughout the profession
Technicians, mechanics	Semi-technical, usually job-related	Moderate	Recognizable symbols already familiar to the reader
General public	Semi-technical or nontechnical	Low	Familiar or obvious symbols only

Readers' Experience. Next consider the amount of experience readers have with the subject of the graphic.

If the Reader Is	Make the Graphic
Novice	Simple, obvious
Expert	Detailed but clearly organized

Language Skills of the Readers. The way and extent to which you use language in graphics depends on the language skills of the reader.

If Readers	Then Design the Graphic to
Read English as a first language	Use English freely (but not where graphics or numbers would be clearer)
Read English as a second language or have difficulty reading	Use English sparingly. Use only common English words. Prefer verbs and concrete nouns. Avoid ambiguous words.
Do not read English	Use no English. Either design the graphic to work without language or translate words to the readers' native language.
Will receive a translated version of the document	Use common, concrete words and simple sentence patterns to promote more accurate translation.

Culture of the Readers. Cultural issues may bar understanding and multilingual graphics alone may not breach these barriers. Consider these barriers when designing graphics.

- **Some readers expect formality:** In some cultures, it is considered insulting to give direct orders. Instructions must be presented as light suggestions. In other cultures, the teacher-student roles (the roles writers and readers of technical documents assume) are separate and absolute. Instructions must be explicit and are expected to be followed literally.
- **Some readers limit the role of writers:** In some oriental cultures, writers are expected only to lay out the facts but not to draw conclusions. Tables and charts, which record data and let the readers draw conclusions, work well in such circumstances.
- **Many readers have sexual taboos that are different from yours:** Sexual mores and roles differ greatly from culture to culture. Consider that European advertisers use nudity more freely than their Madison Avenue counterparts while many Islamic people frown on the depiction of the human body in any form.
- **Some readers assign different symbolic meaning to objects, shapes, and colors:** For instance, the thumbs-up gesture used to hitch a ride and signal OK in the United States is considered an obscene gesture in many Mediterranean countries.

Choose the Right Amount of Graphical Information for the Type of Content You Are Presenting

In addition to considering your audience—its technical specialization, experience, English language skills, and culture—you need to consider the nature of the content you are communicating when determining how much graphical information to present. Each graphic should be designed to communicate a certain type of information and to do so in a simple, clear way. Some characteristics of the presentation depend on the purpose. Consider the following:

If the Purpose of the Graphic Is to	Present Information Like This
Make specific facts accessible from a large amount of data	Concisely organized and presented Easily and efficiently accessed information
Make relationships clear	Showing only major relationships Highlighting trends and primary relationships while suppressing irrelevant details

(continued)

If the Purpose of the Graphic Is to	Present Information Like This
Teach	Free from irrelevant details and distractions Reliant on conventional techniques only Interesting and engaging
Persuade or convince	Attention getting Attractive Credible Focused, making *only* one main point
Attract attention	Stark, focused, simple Dramatic
Label	Quickly recognized, based on familiar elements Distinctive

IMPLEMENTING THE GRAPHICS STRATEGY: TURNING WORDS INTO PICTURES

Many of us are more comfortable with, and proficient at, expressing our thoughts verbally. That's because we're only taught to use words in school and because, in the past, the resources for producing graphics have been expensive. For example, typesetting charts and preparing pictures usually required specialized skills. With high-function word processors, computer graphics programs, and desktop publishing systems, you can produce graphics with little effort—just a lot of thought. This section explains that thought process:

1. How to choose the most appropriate type of graphic for a given type of information
2. How to produce that graphic—even if you don't know how to draw

The next section, *Developing a Repertoire of Graphics Skills: Using Common Graphical Forms*, describes in detail the types of graphics mentioned in this section.

How to Choose the Right Type of Graphic

One of the toughest hurdles facing the communicator is not only thinking in a graphical way but also deciding among the rich variety of established types. Consider the type of information you are trying to communicate: numerical values, logical relationships, procedures and processes, or visual and spatial characteristics.

Numerical Values. Numbers rate, rank, quantify, and describe. The type of graphic best suited to your information depends on whether you want to show exact quantities or only general ratios and on how many numbers are involved.

For This Type of Information	Example of the Type of Information	Use This Type of Graphic
Approximate values		
Absolute	"Customers over 35 years of age were over three times as likely to accept the upgrade as those younger."	Bar or column chart
Proportion	"Expenses were apportioned as follows: Equipment 34%, Maintenance 29%, Supplies 22%, Shipping 10%, and Insurance 5%."	Pie chart, subdivided bar chart, stacked bar chart
Exact values		
Few	"Test results for the forward, mid, and aft compartments were 3.667, 5.224, and 8.482, respectively."	Chart annotated with values, table
Many	Timetable for an airline, mathematical functions	Table
Correlation	". . . and . . . show a high degree of correlation." ". . . is proportional to . . ." ". . . varies inversely as . . ."	Scatter chart with a correlation line
Trend	"Voltage rose steadily from 6 V until leveling off at 14.6 V."	Line chart

Logical Relationships. Graphics can show the logical relationships among objects and concepts.

For This Type of Information	Example of the Type of Information	Use This Type of Graphic
Whole-to-parts relationships	". . . consists of . . ." ". . . comprises . . ." ". . . is composed of . . ." ". . . is made up of . . ."	Exploded parts diagram, organization chart, indented list
Interrelationships	"Model 385 requires Option XB, which in turn requires Upgrades 34C and 34D. Option XD may be substituted for Upgrade 34C, and Model 387 may be substituted for Model 386. However, Model 386 and Upgrade 34D are not compatible."	Network diagram

(continued)

For This Type of Information	Example of the Type of Information	Use This Type of Graphic
Relative importance	Rankings, ratings, test scores	Numbered list, stacked blocks
Decision rules		
Single, simple choice among alternatives	"You can store your document in three formats: as text only, as separate text and graphics, and as integrated text and graphics."	Bulleted list, pictures of alternatives, selection table
Single, complex decision	"If . . . and . . . , then . . . ; otherwise . . . ."	Decision table

Procedures and Processes. Another type of graphic tells you how to do something or how something is made or done.

For This Type of Information	Example of the Type of Information	Use This Type of Graphic
Action sequence		
Performed in a particular order	"First you do . . . . Next you do . . . . Then you do . . . . Finally you do . . . ."	Numbered list, perhaps illustrated with pictures
Performed in any order, but all items required	"Before you begin the procedure, gather these supplies: your program disk, two blank disks, two blank labels."	Checklist of required items
Performed by two people, two machines, or a person and a machine.	First, X does . . . and then Y does. . . . Next X does . . . , and Y does. . . . Finally X does . . . and Y does . . . ."	Playscript (a type of list), table of actions
Decisions		
Series of simple, independent decisions	Diagnostic troubleshooting procedure	Decision tree or table, indented list
Network of simple, interrelated decisions	"If . . . is true, then do . . . . Next, if . . . is true, do . . . ; otherwise do . . . ."	Flowchart

Visual and Spatial Characteristics. More than explaining numerical values, showing logical relationships, or illustrating procedures, many graphics simply show what something looks like or where it is located.

For This Type of Information	Example of This Type of Information	Use This Type of Graphic
Appearance		
Simple subject	An individual, recognizable part	Photograph
Complex subject	Unfamiliar mechanism with many individual components, scene with a distracting background	Line drawing
Spatial relationships		
In 2 dimensions	"The new laboratory is located 3 kilometers south of the test center, just west of Bauxton road."	Map
In 3 dimensions	"The power switch is on the back of the cabinet, near the top-left corner.	Line drawing
Faces		
Specific person	Company president, subject of the document, historical figure	Photograph
Nonspecific person	Everyman, surrogate for reader	Drawing
Internal components		
For general information and a non-technical audience	"The propellant temperature sensor is inside the main feed line, 4 cm upstream from the injector plugs."	Cut away, translucent view
For showing technicians how to assemble an object	"Remove the retaining nut from the shaft, pull off the lock washer, and slide the bearing straight off the shaft."	Exploded parts diagram
For showing professionals how something is designed	Handbook for mechanical engineers	Cross-section

But I Can't Draw!

There you sit, staring at a blank screen or piece of paper. "I can't draw a straight line," you protest. "And if I could, I wouldn't know what to draw. I'm not an artist. I just don't think visually."

Bunk. We all think visually. With practice we can all imagine and create clear, effective visuals. You do need practice though—practice thinking visually and practice with the techniques of drawing.

Practice Thinking Visually. Thinking visually cannot be described; it must be experienced. So experience thinking visually right now:

1. How many windows were in the kitchen of the apartment or house that you lived in when you were nine years old?
2. Imagine a white cube. Imagine that you paint it red on the outside and then cut it into thirds vertically, horizontally, and longitudinally as shown, resulting in 27 small cubes.

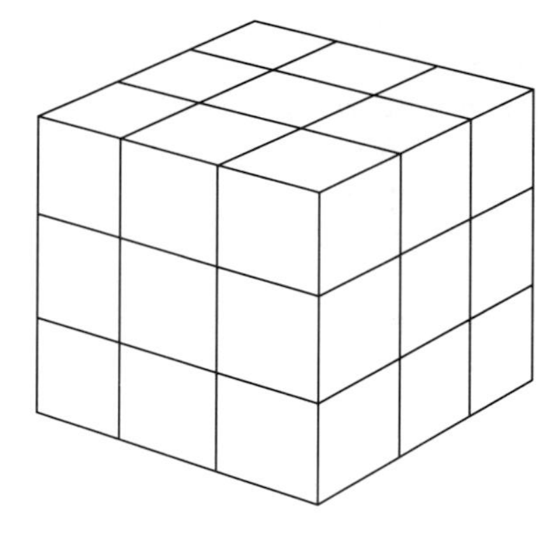

   a. How many of these small cubes are red on three sides?
   b. On only two sides?
   c. On only one side?
   d. How many are unpainted?

How did you answer these questions? By forming visual images. While you're at it, try these additional exercises in visual thinking. Try imagining each of the following:

1. A kettle of soup heating and then boiling over
2. Going to work in a helicopter
3. A charging elephant turning into a zebra and then into a lion
4. A waterfall flowing backwards and eventually running dry
5. A personal computer turning to stone

Tricks and Tips for Learning to Draw. If you are not a professional illustrator, you may not know the many techniques used to simplify and streamline the task of drawing. Some of these techniques are conceptual and some are pragmatic, but they all can increase your drawing abilities and confidence. Consider these techniques:

- Start with a calm, observant state of mind. Take a few slow, deep, deliberate breaths to steady your thoughts and hands.
- Trace the original. Tracing helps you develop a feel for the activity of sketching the shapes and curves of objects. And, for many purposes, the tracing may be all you need. Never, however, publish a tracing of copyrighted artwork without permission.
- Instead of drawing the object itself, draw the empty spaces within and around the object. This is called sketching negative space.
- Stand the original on its head. This will encourage you to draw the lines and surfaces presented to your eye, rather than your memories and interpretations of what you are seeing.
- Learn to draw basic geometric shapes, such as a block, sphere, cylinder, and cone. Then draw real objects as if made up of just these simple shapes.
- Learn to draw other basic shapes and common objects. Develop simple drawings of the human face, frame, and hands.
- Measure the scene through a transparent grid. Draw one small square of the grid at a time. Use the grid to measure relative lengths and sizes of edges and surfaces.

DEVELOPING A REPERTOIRE OF GRAPHICAL SKILLS: USING COMMON GRAPHICAL FORMS

Previous sections in this chapter stressed the importance of planning a graphics strategy and determining which words to turn into pictures before actually preparing your graphics. With plans in hand, you can begin. Preparing graphics first requires a familiarity with the various types of graphics and then with the techniques for improving their quality and effectiveness. This section explores the various type of graphics: what they are and when to use them. Specifically, this section explores these types of graphics starting with text-based graphics and ending with picture-based graphics:

- Lists
- Tables
- Visual symbols and icons
- Photographs
- Drawings
- Charts

- Diagrams
- Maps

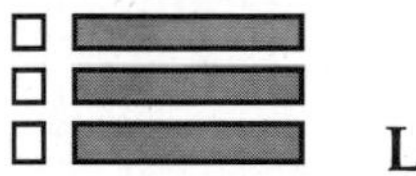

Lists

Lists are a visual means of presenting text. Lists separate individual items of text and present them in sequential order.

When to use lists:

- To present a group of related items of the same class
- When the items naturally fall in a series
- When the order of the items is meaningful

Rules of thumb:

- Select items that the audience will recognize.
- Select items of the same class. If the class is not obvious, tell the audience what the items have in common.
- Complete the list. Do not leave gaps or omit important items.
- Order items to show the crucial relationship among them. Use
    - **Spatial** order for describing visual appearance
    - **Chronological** order for describing processes and spelling out procedures
    - **Logical** order for explaining reasoning and presenting arguments. Logical order can argue from cause to effect or trace reasoning from result to its cause.
    - **Rhetorical** order to enhance the impact of a list or to draw attention to one particular member.
    - **Indexed** order when the list is long and the reader must scan it to find a particular item. (Examples of indexed order include alphabetical or numeric order.)
    - **Ranked** order to show relative importance, power, or rating.
- Begin each list item with the same type of word, such as a noun or verb. For example, if the first word of the first list item is an adjective, then the first word of each list item should be an adjective.
- Present the list in the form that best suits the purpose of the list and the available space. Use the following guidelines:

Type	When to Use It	Example
Sentence lists	Where space is tight	Colors observed were: ▪ red ▪ yellow ▪ green ▪ blue
Bulleted lists	To show alternatives, especially when there is no natural order among the items and not all items are necessary	Colors observed were ▪ red ▪ yellow ▪ green ▪ blue
Checklist	For readers to select among listed items, especially where all items are required	Before you begin, gather the tools you will need: √ 14 mm crescent wrench √ flat-head screwdriver √ micrometer
Numbered lists	For items that must be considered in a particular order, for instance to indicate a ranking or to present step-by-step instructions	To process the film: 1. Load the canister. 2. Turn on the processor. 3. Wait 5 minutes.
Grades	To divide a range of values into labeled segments	A B C D E
Rulers	To show where something ranks along a continuous numerical scale	A B C D E
Ladders	To divide an abstract, continuous scale into definite categories, each of which builds on or adds to the previous one, for instance, levels in the food chain	A B C D E
Playscript	To list the actions in a multiperson procedure using a dialog format	A: Start machine. B: Feed in casting. A: Take out casting. B: Stop machine.

Indented lists	To show secondary hierarchical relationships among items of a series	Main item
		   Sub item
		   Sub item
		Main item

- Separate the items of the list visually and make it easy for the eye to scan the list.

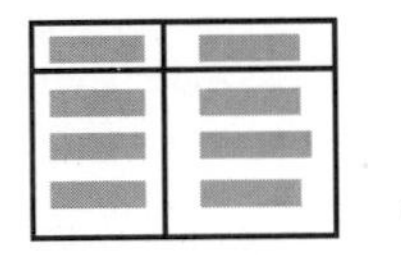

Tables

Tables arrange many individual pieces of information, usually text or numbers, in neat, accessible rows and columns.

When to use tables:

- To present a large amount of detailed information in a small space.
- To facilitate detailed, item-to-item comparisons.
- To precisely show individual data values.
- To simplify access to individual data values.

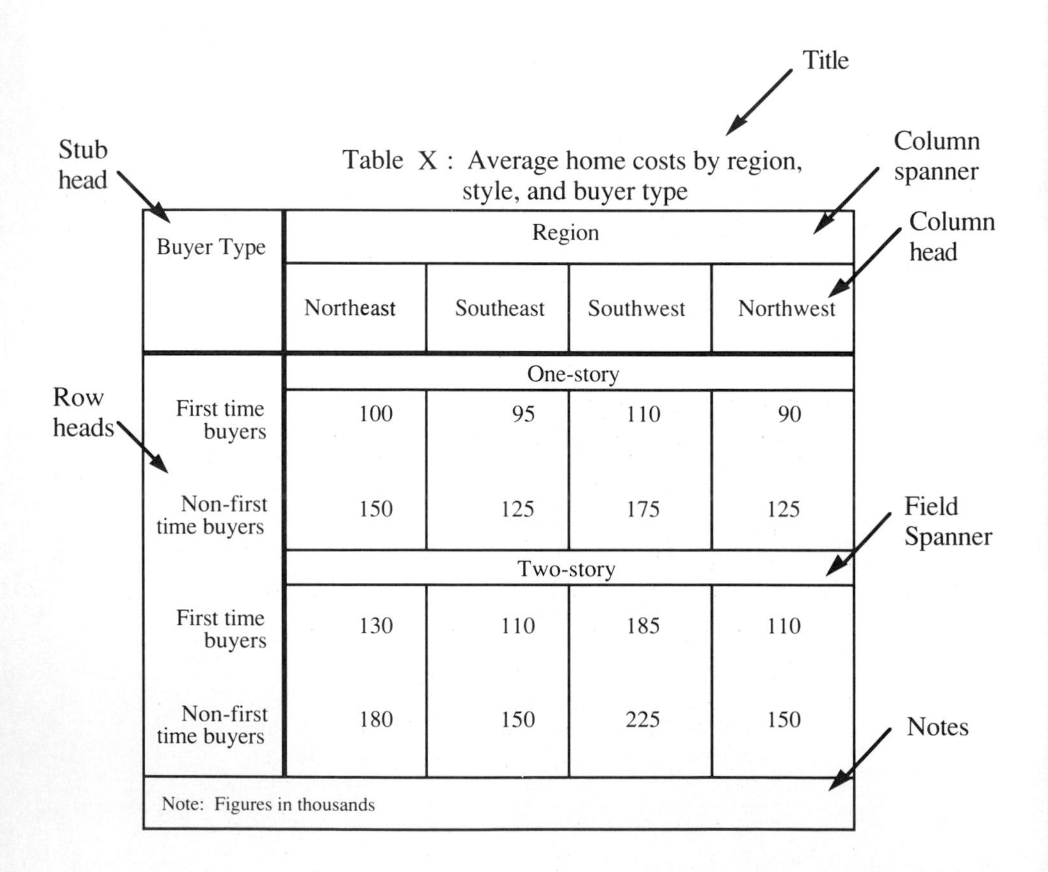

Table X : Average home costs by region, style, and buyer type

Buyer Type	Region			
	Northeast	Southeast	Southwest	Northwest
	One-story			
First time buyers	100	95	110	90
Non-first time buyers	150	125	175	125
	Two-story			
First time buyers	130	110	185	110
Non-first time buyers	180	150	225	150

Note: Figures in thousands

Rules of thumb:

- Minimize the number of lines in a table, except where necessary to show the scope of column spanners (labels for multiple columns) or to separate closely spaced columns or rows.
- Design the table for reliable, quick scanning. Order rows and columns so that more common entries are toward the top and left where the eye is most likely to look. Beware of widely spaced columns.
- Include numbers, text, and graphics, as appropriate, in the cells, or information spaces, of the table.
- Make tables complete. Give each table a title, complete stub (left-most column) and column heads, and notes so that the table is effective, even without surrounding text.
- Avoid mixing different types of information in a column. Apply the same format to all information in a column. For instance, include the same number of decimal points for all numbers in a column.

Visual Symbols and Icons

Visual symbols are simple, sometimes abstracted, visual representations—or pictures—of objects, actions, and concepts. Examples include these, which are used as signposts for the various types of standard graphics:

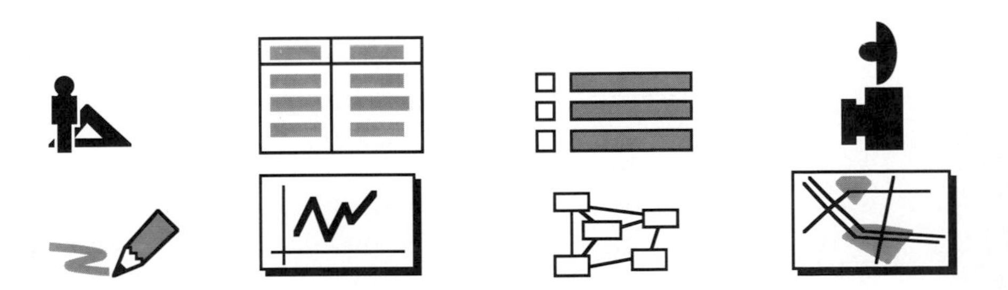

When to Use Visual Symbols. Use visual symbols where you would use words or phrases:

- To label objects.
- To jog the readers' memories.

- For quick recognition.
- For multilingual signs and labels, such as those found in airports and public buildings.

Rules of thumb:

- Use visual symbols with words, not as substitutes for words. Words and pictures work better together than alone.
- Design visual symbols for use in a specific context. Combine symbols with information available in the context so the meaning is clear.
- Beware of cultural differences in the meaning of symbols. For instance, the thumbs-up gesture, used to mean OK and to hitch a ride in the U.S., is an obscene gesture in many Mediterranean countries.
- Test visual symbols with actual readers under actual viewing conditions.

Photographs

Photographs are recorded representations of reality.

When to use photographs:

- To show specific people. Use photographs when the subject is recognizable or you want to make the subject recognizable. Do not use photographs where individual identity is not important, as in work procedures, where the person shown merely stands in for the reader.
- To show human faces, hands, and other objects difficult to accurately draw. Photos of the human face are especially effective in conveying emotion.
- To show actual appearances where details, color, tones, and textures are important. Do not use photographs if the background is cluttered or distracting.
- To prove that something is real.

Rules of thumb:

- Simplify the photograph. Remove distracting details. Zoom in or trim the photograph to cut out irrelevant objects.
- Direct attention to the subject. Light the subject well and focus on it. Narrow the depth of field to throw nearer and farther objects out of focus.

- Clearly indicate the scale and viewpoint. Orient the viewer. Include a familiar object in the scene.
- Avoid dimly lit or low-contrast photographs of three-dimensional objects, especially the interior of machinery, cabinets, and undistinguished spaces.
- Capture depth in photographs. Use lighting, viewpoint, and focus to emphasize the three-dimensional nature of the subject. Avoid flatly lit, straight-on mug shots.

Drawings

Drawings use lines and shading to show the physical appearance and spatial relationships of an object or scene.

When to use drawings:

- To show the appearance of an object or scene but not unnecessary detail.
- To show objects that do not exist yet.
- To show views impossible to create without destroying the subject in the process.
- To show nonspecific people, suppressing personal or idiosyncratic details.
- To include a person in a scene without having the person dominate the scene.

Rules of thumb:

- Keep drawings simple. Avoid details that do not contribute to the point you are trying to make.
- Emphasize parts of the drawing that carry the most information, such as corners, curves, and outlines.
- Select the type of drawing to match the audience, for instance, cutaways for the general public, exploded views for mechanics, and cross-sections for engineers.
- Use cartoons only where a light touch is appropriate.

Charts

Charts represent relationships among data, such as distances and proportions. (The term "charts" as used here includes graphs and similar displays of quantitative information.)

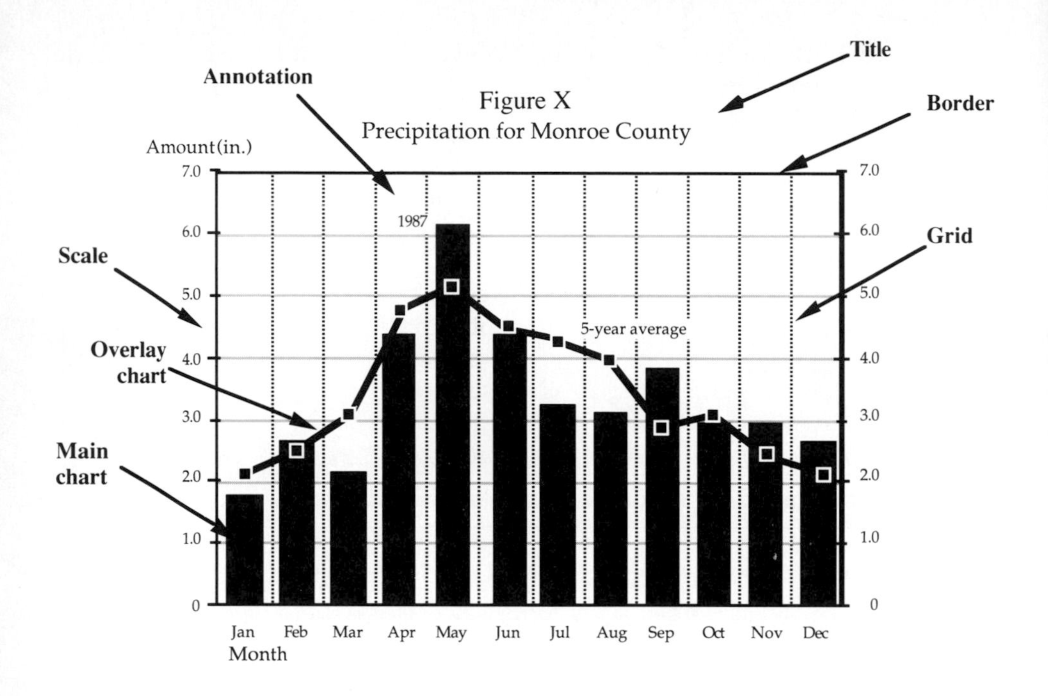

When to use charts: Use charts to

- Emphasize trends, relationships, and patterns in the data to convincingly make a point with the reader.
- Forecast and predict future values. Charts are superior to tables or text when the reader must interpolate or extrapolate from given values.
- Present large amounts of complex data without overwhelming the reader's comprehension.
- Add credibility. Many people feel graphics are more authentic and truthful than words or numbers alone.
- Interest the reader in the data. Well prepared charts can intrigue and seduce the casual browser into reading the text.

Do not use charts to

- "Jazz up" or "spritz up" boring numerical data. If the numbers are boring, the chart will be boring, too.
- "Break up" text. Instead, integrate the chart seamlessly into the flow of thought.
- Show precise numbers. Use a table instead of a chart when you must provide supporting numbers and allow detailed comparisons.
- Communicate to an audience unfamiliar with charting techniques.
- Interpret, explain, evaluate, and review data. Use text instead.

Rules of thumb:

- Keep the chart simple. Avoid decorative borders and backgrounds.
- Focus on the data. Make the data the primary information in the chart and display it so that trends stand out.
- Avoid misleading charts that result from a nonobvious scaling of values (such as logarithmic or geometric scales in charts for non-technical readers).
- Make charts complete. Use titles, notes, and annotations to create self-contained charts that do not require reading surrounding text.
- Use similar symbols, textures, or colors for related items of data. Likewise, use different coding schemes to distinguish separate data elements or series.
- Use curves and lines to show trends and to connect data points.

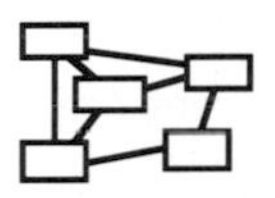

Diagrams

Diagrams are visual tools that use a network of connected symbols to show abstract relationships among objects or concepts.

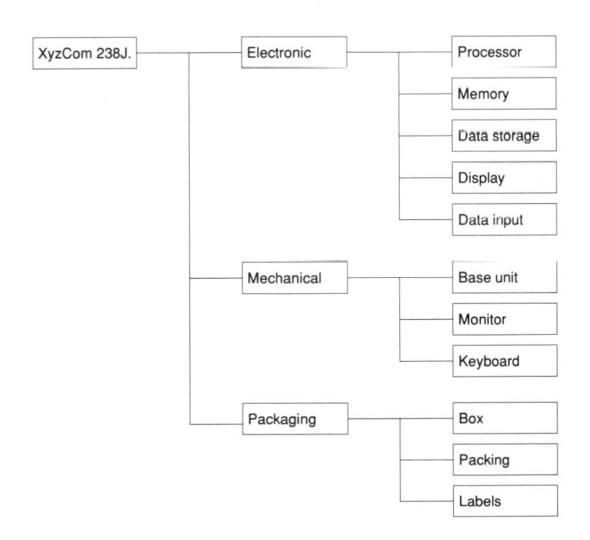

When to use diagrams:

- To show how parts are organized into a coherent system.
- To record the interrelationships among many components of a system.
- To communicate a pattern of relationships.

Rules of thumb:

- Keep the diagram simple. Focus attention on the relationships shown.
- Avoid complex, abstract technical symbologies in diagrams that your audience will not understand.
- Use established forms and conventions, especially in diagrams for specific technical disciplines.
- Label all symbols and links between them. Provide a key for symbol shapes, colors, and line styles.
- Emphasize the primary pattern of organization, such as a sequence, hierarchy, or grid.

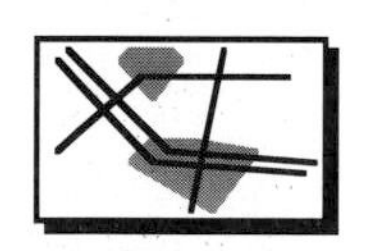

Maps

Maps are special types of diagrams that show the location of objects in space and their relationships.

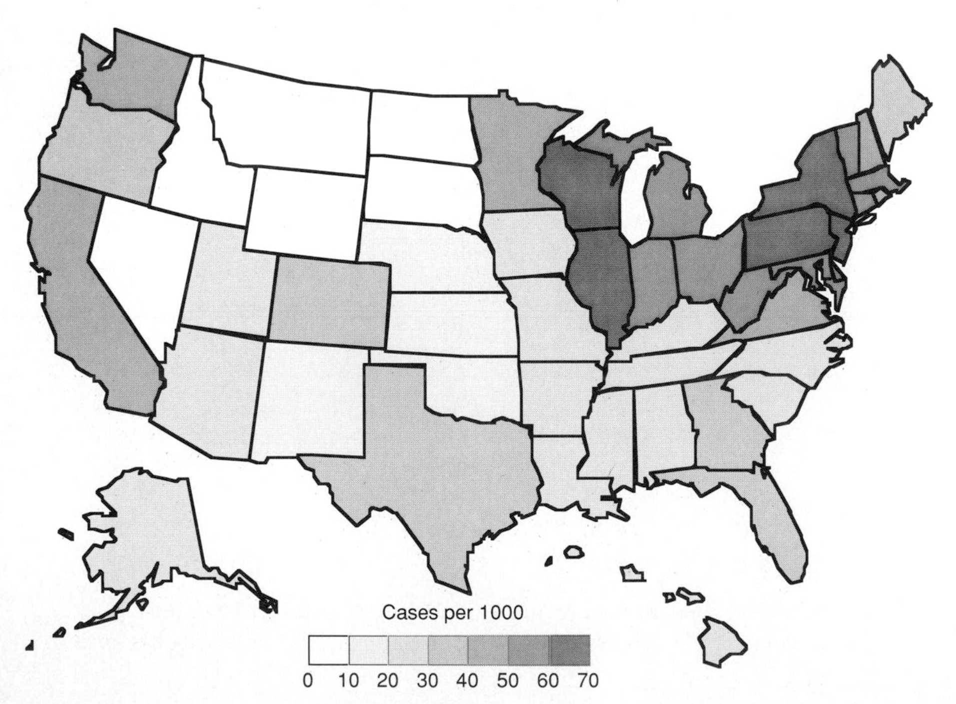

When to use maps:

- To show where things are located in relation to each other and to the reader's current location.
- To show geographical distribution of data or objects.
- To help readers traverse a complex system or terrain.

Rules of thumb:

- Keep maps simple. Design the map for a single purpose and only display details necessary for that purpose.
- Use obvious or familiar symbols, especially in maps for the general public. In maps for specialists, use symbols already well established in the specialist's domain.
- Follow natural color conventions. For example, use blue to indicate bodies of water, not red.
- Simplify the display. For instance, straighten out roads if doing so does not confuse crucial relationships.
- Include a scale and a full legend.

DEVELOPING A REPERTOIRE OF GRAPHICAL SKILLS: INCREASING THE QUALITY AND EFFECTIVENESS OF YOUR GRAPHICS

In addition to properly preparing the various types of graphics, you can follow some general strategies to ensure the quality and effectiveness of your graphics. This section presents those strategies:

1. Linking and separating
2. Using depth cues
3. Using graphical dynamics
4. Using color
5. Fitting graphics to the medium

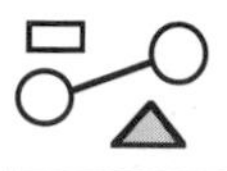

Linking and Separating

Linking and *separating* are techniques to make objects separate from others in a graphic and show how objects are related. Here are three techniques for linking and separating:

To make something stand out from the background:

- Make it a different brightness, color, or texture.
- Draw it at a different depth. Let it cast a shadow on the background.
- Emphasize its solidity and depth.

To make something stand out from similar objects:

- Make it larger or smaller.
- Make it lighter or darker.
- Make it a different color.
- Give it a different shape.
- Surround it with blank space or a box.
- Let it violate a pattern established by the others.

To link separate objects:

- Draw them with the same brightness, color, or texture.
- Arrange them in a simple geometric pattern.
- Cluster them together.
- Connect them with lines.

Using Depth Cues

Depth cues are a means of showing three dimensions in graphics. Use depth cues

- To show depth and distance.
- To depict solid objects.
- To highlight important objects. To make them stand out.
- To add interest.

Just as you have a variety of graphic types to use, you also have a variety of depth cues to use. The following chart presents these cues.

Depth Cue	Where and How to Use It	Example
Overlap	Nearer objects block our view of farther objects.	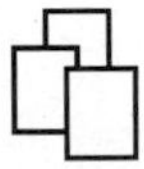
Scale	Nearer objects appear larger than more distant objects.	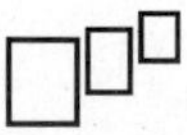

(continued)

Elevation	Farther objects appear higher in our field view.	
Shade and shadow	Solid objects are darker on the side away from a light source. Opaque objects cast shadows in the direction opposite the light source.	
Linear perspective	Parallel lines appear to converge to an infinitely distant vanishing point.	
Aerial perspective	Distant objects appear paler, bluer, and less distinct than nearby objects.	
Color perspective	Colors from the red end of the spectrum appear nearer than ones from the blue end.	
Gradient of texture	Toward the horizon, lines seem to get closer and closer together and textures grow finer.	
Selective focus	At any one time, only objects in a certain range of distances are in sharp focus. Nearer or farther objects fall progressively out of focus.	

Using Graphical Dynamics

Graphical dynamics are artistic conventions for conveying movement and change on the static display. Use graphical dynamics:

- To suggest movement in static displays
- To show change
- To direct the reader's eye across the display
- To draw attention

Several types of graphical dynamics are available to you. The following chart presents these graphical dynamics:

Graphical Dynamic	Where and How to Use It	Example
Arrows	Arrows point the direction of motion. Use them to indicate motion and show its direction.	
Pregnant moment	Show objects in an unstable position, typical of the beginning of a motion.	

(continued)

Graphical Dynamic	Where and How to Use It	Example
Blurred edges	Make the edges of speeding objects indistinct.	
Speed lines	Draw speed lines trailing behind the object. Use the number, length, and direction of the lines to indicate the object's approximate velocity.	
Shake lines	Draw faint lines parallel to the edges of an object to show vibration, turbulence, or nervousness.	
Ghost images	Draw faint images of an object trailing behind the object.	
Sequence of snapshots	Show the object at various stages of change.	
Freeze-frame photographs	Show an instantaneous snapshot of a rapidly moving object. Paradoxically removing all motion from the scene actually emphasizes the motion.	
Energetic brush strokes	Use a turbulent or vibrant texture to convey a sense of excitement or turmoil.	

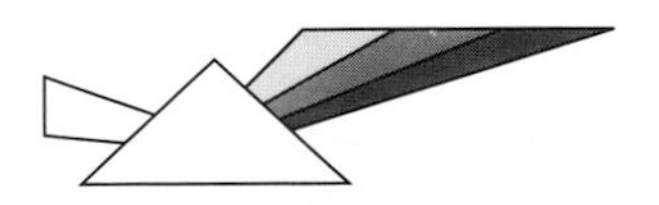

Using Color

Color has proven to be an effective attention getter in all types of documents, although color has not proven to be an effective aid in readability. Use color to:

- Link objects and different colors to distinguish objects.
- Focus attention on:
  - Small but important objects
  - Headings and other important labels
  - Warnings, cautions, and notes
  - Actions the reader is supposed to take

- Speed search, for instance, by color-coding the different symbols in a complex diagram
- Express a range of values, for instance, temperatures on a weather map (red = hot, blue = cold)

Rules of thumb:

- Limit the number of colors. Use only four to six color codes.
- Make colors distinct.
- Use natural color associations and color conventions from the reader's world of work: for instance, red for danger, blue for water, green for vegetation
- Pick harmonious and pleasant colors. Balance colors. Avoid garish and conflicting colors. Tastes vary, so test your choices with representative readers.
- Don't use colors that compromise legibility. Maintain contrast between foreground and background.
- Do not depend on color alone for any critical distinction.

Fitting Graphics to the Medium

The guidelines presented in this chapter for developing graphics can be used no matter what medium you use. But each medium does have its own characteristics, and you should consider these characteristics as you develop your graphics. The following chart discusses these characteristics.

Medium	Characteristic	Special Consideration
Book	Pace controlled by reader	Use as much detail as necessary
	Color printing is expensive and complex	Avoid color for decoration. Use color only where necessary to communicate.
	Printed at high resolution	Make the minimum text size 6–8 points (10 points if reproduction is poor or lighting is dim)
TV	Color used widely	Use color
	Displayed at low resolution	Use only very simple or recognizable images Make the height of text at least 1/20 height of screen
	Audience is nontechnical	Use only common types of graphics
	Width:height ratio is 4:3	Proportion graphics to fit

(continued)

Medium	Characteristic	Special Consideration
Computer screen	May have color	Use color if available
	Displayed at low resolution	Use only simple graphics Only use simple type faces
Slide	Shown in darkened room	Light subject on dark background
	Medium resolution	Make the height of text at least 1/50 height of the graphic
	Color is used widely	Use color
	Width:height ratio is 4:3	Proportion accordingly
	Paced by presenter	Keep graphics simple. Present main points or trends, not details.
Viewgraph, overhead transparency	Shown in lighted room	Dark subjects on light backgrounds
	Medium resolution	Make the height of text at least 1/50 height of the graphic
	Paced by presenter	Keep the graphic simple, showing main points and trends only
	Width:height ratio is 3:5 or 5:3	Proportion accordingly

CHECKING GRAPHICS

Once the graphic is complete, you must check it against your plan to ensure that it will indeed accomplish its purpose. Points to check include:

- √ Is your primary idea clear and immediately obvious?
- √ Do all details support the main idea?
- √ Does the subject matter stand out from background and annotation?
- √ Is this the most appropriate type of graphic (table, chart, diagram, etc.) for the purpose?
- √ Will skeptical viewers trust the graphic?
- √ Is every point, line, symbol, and word necessary?
- √ Is the graphic legible under actual viewing conditions?
- √ Can you reproduce or display it clearly, economically, and reliably?
- √ Does the graphic follow conventions familiar to viewers?
- √ Is the graphic consistent with the text and other graphics?
- √ Is the graphic pleasing to look at?

SUMMARY

This chapter explained how to present information visually. You should present information visually—with graphics—because:

- Graphics communicate what words cannot.
- Graphics are understood more quickly than words.
- Concrete images are remembered better than abstract concepts.
- Graphics entice readers.
- Thoughts are visual—even in words.

You begin using graphics by analyzing, planning, and selecting a strategy that includes the following steps:

1. Choosing the right amount of graphical information for your audience
2. Choosing the right amount of graphical information for the type of content

Next, you implement the strategy by choosing the right type of graphic. Common graphical forms that you can use include

- Lists
- Tables
- Visual symbols and icons
- Photographs
- Drawings
- Charts
- Diagrams
- Maps

You can increase the quality and effectiveness of your graphics by using

- Lining and separating
- Depth cues
- Graphical dynamics
- Color
- Graphics that fit the medium

LIST OF TERMS

charts Tools for representing relationships among data, such as distances and proportions.

depth cues Means of showing three dimensions in graphics.

diagrams Visual tools that use a network of connected symbols to show abstract relationships among objects or concepts.
drawings Use of lines and shading to show the physical appearance and spatial relationships of an object or scene.
graphical dynamics Artistic conventions for conveying movement and changes on a static display.
graphics Visual formats for presenting information.
icons See *visual symbols*.
linking and separating Technique to make objects separate from others in a graphic and show how objects are related.
lists Visual means of presenting text. Lists separate items of text and present them in sequential order.
maps Special types of diagrams that show the location of objects in space and their relationships.
photographs Recorded representations of reality.
tables Arrangements of many individual pieces of information, usually text or numbers, in neat, accessible rows and columns.
visual symbols Simple, sometimes abstracted, visual representations—or pictures—of objects, actions, and concepts. Also called *icons*.

SUGGESTED READING

Adams, A. (1969). *Artificial-light photography.* Hastings-on-Hudson, NY: Morgan & Morgan.

Adams, A. (1969). *Natural-light photography.* Hastings-on-Hudson, NY: Morgan & Morgan.

Adams, A. (1970). *Camera and lens: The creative approach.* Hastings-on-Hudson, NY: Morgan & Morgan.

Adams, J. L. (1979). *Conceptual blockbusting: A guide to better ideas.* New York: W. W. Norton.

Arnheim, Rudolf. (1979). *Art and Visual Perception: A psychology of the creative eye.* Berkeley, CA: University of California Press.

Arnheim, R. (1969). *Visual thinking.* Berkeley, CA: University of California Press.

Barratt, K. (1980). *Logic and design in art, science, and mathematics.* New York: Design Press.

Bertin, J. (1983). *Semiology of graphics.* Green Bay, WI: University of Wisconsin.

Buehler, M. F. (1977). Report construction: tables. *IEEE Transactions on Professional Communication, 20* (1), 29–32.

Buehler, M. F. (1980). Table design—When the writer/editor communicates graphically. In *Proceedings of 27th International Technical Communication Conference.* Minneapolis: Society for Technical Communication, pp. G 69–73.

Carliner, S. (1987). Lists: The ultimate organizer for engineering writing. *IEEE Transactions on Professional Communication, 30* (4), 218.

Christ, R. E. (1975). Review and analysis of color-coding research for visual displays. *Human Factors, 17* (6), 542–570.

Dalton, S. (1983). *Split second: The world of high-speed photography.* Salem, NH: Salem House.

De Grandis, L. (1986). *Theory and use of color.* New York: Abrams.

Dreyfuss, H. (1984). *Symbol sourcebook: An authoritative guide to international graphic symbols.* New York: Van Nostrand Reinhold.

Edwards, B. (1979). *Drawing on the right side of the brain: A course in enhancing creativity and artistic confidence.* Los Angeles: J. P. Tarcher.

Feininger, A. (1970). *Total picture control.* New York: Chilton Books.

Ford, D. F. (1984). Packaging problem prose. In *Proceedings of 31st International Technical Communication Conference.* Seattle: Society for Technical Communication, pp. WE 52–55.

Frutiger, A. (1989). *Signs and Symbols: Their design and meaning.* New York: Van Nostrand Reinhold.

Gettys, D. (1986). If you write documentation, then try a decision table. *IEEE Transactions on Professional Communication, 29* (4), 61–64.

Gombrich, E. H. (1969). *Art and illusion: A study in the psychology of pictorial representation.* Princeton: Princeton University Press.

Gregory, R. L. (1980). *Eye and brain: The psychology of seeing.* New York: Oxford.

Hamm, J. (1967). *Cartooning the head and figure.* New York: Grosset & Dunlap.

Hanks, K., & Belliston, L. (1977). *Draw! A visual approach to thinking, learning, and communicating.* Los Altos, CA: William Kaufmann.

Hayes, J. R. (1981). *The complete problem solver.* Philadelphia: The Franklin Institute Press.

Herdeg, W. (Ed). (1976). *Graphis diagrams.* Zurich: The Graphis Press.

Hill, M., & Cochran, W. (1977). *Into print: A practical guide to writing, illustrating, and publishing.* Los Altos, CA: William Kaufmann, Inc.

Horton, William. (1983). Toward the four-dimensional page. In *Proceedings of 30th International Technical Communication Conference.* St. Louis: Society for Technical Communication, pp. RET 83–86.

Horton, W. (1985). Quick-relief documentation. In *Proceedings of 32nd International Technical Communication Conference.* Houston: Society for Technical Communication, pp. WE 4–7.

Horton, W. (1991). *Illustrating computer documentation: The art of presenting information graphically on paper and online.* New York: John Wiley.

Horton, W. (1991). *Secrets of user-seductive documents: Wooing and winning the reluctant reader.* Arlington, VA: Society for Technical Communication.

Huff, D. (1954). *How to lie with statistics.* New York: W. W. Norton & Company.

Human Factors Society. (1988). *American national standard for human factors engineering of visual display terminal workstations* ANSI/HFS 100-1988. Human Factors Society.

Jarvenpaa, S. L., & Dickson, G. W. (1988). Graphics and managerial decision making: Research-based guidelines. *Communications of the ACM, 31* (6), 764–774.

Kaufman, L. (1979). *Perception: The world transformed.* New York: Oxford University Press.

Kodak. (1989). *Professional photographic illustration.* Rochester, NY: Eastman Kodak Company.

Lefferts, R. (1982). *How to prepare charts and graphs for effective reports.* New York: Barnes & Noble.

Lockwood, A. (1969). *Diagrams.* New York: Watson-Guptill.

MacGregor, A. J. (1979). *Graphics simplified: How to plan and prepare effective charts, graphs, illustrations, and other visual aids.* Toronto: University of Toronto Press.

Mann, G. A. (1984). How to present tabular information badly. In *Proceedings of 31st International Technical Communication Conference.* Seattle: Society for Technical Communication, pp. SE 48–51.

Martin, J., & McClure, C. (1984). *Diagramming techniques for analysts and programmers.* Englewood Cliffs, NJ: Prentice-Hall.

McKim, R. H. (1980). *Thinking visually: A strategy manual for problem solving.* Belmont, CA: Lifetime Learning Publications.

Modley, R. (1976). *Handbook of pictorial symbols.* New York: Dover.

Nelms, H. (1981). *Thinking with a pencil.* Berkeley, CA: Ten Speed Press.

Ogilvy, D. (1985). *Ogilvy on advertising.* New York: Vintage Books.

Robertson, B. (1988). *How to draw charts and diagrams.* Cincinnati, OH: North Light Books.

Schmid, Calvin F., and Schmid, E., *Handbook of Graphic Presentation.* New York: John Wiley & Sons, 1979.

Selby, P. H. (1979). *Using graphs and tables.* New York: John Wiley & Sons.

Tufte, E. R. (1983). *The visual display of quantitative information.* Cheshire, CT: Graphics Press.

Tufte, E. R. (1990). *Envisioning information.* Cheshire, CT: Graphics Press.

White, J. V. (1988). *Graphic design for the electronic age.* New York: Watson-Guptill.

White, J. V. (1990). *Color for the electronic age.* New York: Watson-Guptill.

Winn, W., & Holiday, W. (1982). Design principles for diagrams and charts. In *The technology of text: Principles for structuring, designing, and displaying text.* Englewood Cliffs, NJ: Educational Technology Publications, 277–299.

Winn, W. (1990). Encoding and retrieval of information in maps and diagrams. *IEEE Transactions on Professional Communication,* 33 (3), 103–107.

CHAPTER 7

DESIGN THAT DELIVERS—FORMATTING INFORMATION FOR PRINT AND ONLINE DOCUMENTS

Martha Andrews Nord

Vanderbilt University

Beth Tanner

Tanner Corporate Services

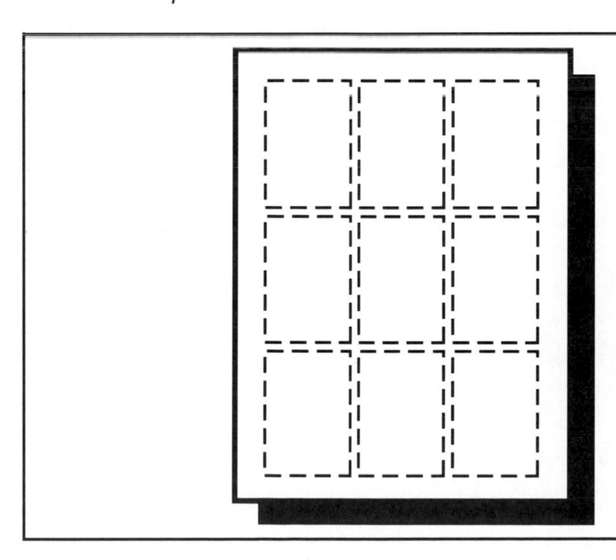

> "How can we design text that both appeals to an intended readership and enhances their ability to understand, learn, use and retrieve information?"
>
> (Schriver, 1989)

Schriver poses the question that suggests the focus of current efforts by researchers and practitioners in document design. Both groups emphasize that the driving force behind all design decisions is the effect of communication on the performance of readers in learning, using, and retrieving information.

Design decisions do not come as an afterthought at the production phase, a way to make a document attractive after the writers have done the "real work." As soon as you have determined who the readers are and how they will use the information, you are ready to make design decisions. Only the implementation of design choices occurs at the production phase.

You can influence readers' responses by using both verbal and visual strategies. In fact, the appropriate mixture of visual elements and verbal techniques strengthens communication. Although communicators have no absolute rules determining an optimal design, a growing body of research is expanding our understanding of how visual and verbal features influence readers' interaction with a text and has confirmed notions of what is good document design (Benson, 1989).

By learning from the research, technical communicators can make informed decisions about combining visual and verbal features that significantly increase the usefulness of a text. To use any of the techniques and tools effectively, however, you must understand your audience, purpose, and each tool.

Chapters 5 and 9 explain strategies for presenting information verbally. This chapter describes the *visual* tools you can use to influence readers' performance, whether the information is on a page or screen. The chapter complements Chapter 6, which discusses illustrations, a crucial visual element. Specifically, this chapter presents the results that design features can help readers achieve and explores techniques and tools that you can use to achieve these results. The chapter is organized in two parts:

- Reader results—the responses that design features can help readers achieve
- Visual tools
    - "Big-picture" techniques for designing entire documents and sections within documents
    - "Little-picture" tools for designing pages and screens

RESULTS READERS NEED TO ACHIEVE

Whatever the form of technical communication, readers need to see and read the information with ease, to find specific parts of it, to understand it, to remember it, and even to enjoy it. The following chart explains each of these results.

Reader Result	**Description**
Ease of reading	Readers' ability to see and read the information.

Ease and speed of reading are determined by the legibility of text and the graphic presentation. Legibility and presentation are influenced by four design choices:

1. Layout (the location of information on a page or screen) and the amount of white space
2. Type size and style of letters
3. Spacing between letters and lines
4. Contrast, or the sharpness of characters against the surface on which they are written

In printed material, for example, ease of reading is influenced by upper- and lowercase type. Paragraphs in any *font* typed all uppercase can slow reading speed. It is much easier for most people to read the mixture of upper- and lowercase because the shapes of the letters help present the words as a unit.

TECHNICAL COMMUNICATION

technical communication

Highly stylized characters used to illustrate an old manuscript challenge readers even more than type that is all uppercase. Imagine trying to read a whole paragraph of these characters:

𝔄 𝔅 ℭ 𝔇 𝔈 𝔉 𝔊

The ease of reading information on a display is influenced by the physical characteristics of the screen. For example, you must achieve legibility within a space that is smaller than a printed page and has a different shape. Display screens are brighter than pages, so achieving a readable contrast between the text and background is challenging.

Visual tools you can use to improve reading ease include

- Placement or layout
- White space
- Typography

Accessibility

Readers' ability to find a specific piece of information in a document or program.

Research shows that the success of a document depends on how useful it is for the intended readers. As noted in Chapter 1, the way people read has two important implications for design:

1. Readers need to recognize information already known so they can easily find the unknown.
2. Readers benefit from schemata, or familiar patterns, that set up expectations to make documents predictable.

Imagine learning a new word processing package with no cross references to match new terms with the ones you already know, or trying to find your eligibility for benefits in an employee handbook with no table of contents.

Visual tools can be aids for retrieving information (called *retrieval aids*) and for establishing predictable patterns to help readers locate information. For printed material, tools that help readers negotiate the text are based on mental models of books (Shirk, 1988). These aids include

- Tables of contents
- Indexes
- Headings
- Highlighting

Tools based on books may not transfer to online documents and multimedia presentations. Without a model for navigating through these types of documents, readers can easily get lost in midprocess. Visual cues to help readers navigate online information and retrieve details from it include

- Menus, or lists of options
- Icons, or visual symbols for commands
- Consistent screen formats or layout
- Highlighting with color, boxes, type change, blinking
- Cues and the like

Comprehension Readers' ability to understand information.

Experts may have no trouble comprehending a relatively dense layout (a page covered by text unbroken by frequent paragraph divisions or headings), but novice readers need to read information in smaller units. It is easy to imagine a lawyer's comfort level with unbroken text in a legal brief or a beginning student's anxiety when first seeing small print and dense pages in a new textbook.

The literacy level of the audience also influences design choices. Pictures, for example, aid comprehension, especially with readers who have poor reading skills. Pictures of food on cash register keys speed entry and improve accuracy. For a more specific example, a social service agency in Ohio decided on posters rather than brochures to communicate good parenting techniques to an audience with little understanding of parenting and a low reading level. Using large pictures and simple text aided comprehension (Floreak, 1989).

Helping readers comprehend online documents presents challenges similar to those raised by accessibility. Just the effort to concentrate on where they are in a series of online tasks can interfere with readers' comprehension of the material. Retrieval aids can enhance comprehension if they are complete and simple enough for the novice, but easy for the expert to bypass.

The small size of the screen also influences comprehension because it limits the amount of information that readers receive. Selecting what to include within space limitations becomes a challenge for the writer. The small space may, however, become an asset for readers if it forces writers to organize information in easy-to-digest units.

Retention Readers' ability to recall both information and its location in a document.

You must carefully control the amount and complexity of information you present to avoid overloading readers' memories. Whether reviewing current research in a specialty area, going through a new manual, or simply sorting a day's mail, readers can quickly reach the saturation point defined by "Miller's magic number," a proven number of items that the brain can process: 7 ± 2 (Miller, 1956).

Design features can also promote retention by providing visual cues and familiar patterns and images. In printed materials, layout, icons, and typography create emphasis and repeated patterns for memory aids. Repeated visual cues, such as the same placement of headings, use of color, and type for all manuals in a series, give readers what Rubens and Rubens (1988) call a "security blanket" to depend on no matter where they are in a document.

Using familiar symbols can also aid memory. For example, readers easily recognize an octagonal red sign to signal STOP or a skull and crossbones to warn of poison. Using existing symbols or creating new ones can help readers remember important information and help them recognize where they are in a document.

Design features can also promote retention in online documents. To help readers remember their location in a document, establish specific keys to take readers to the end or the beginning of the text or to help screens. For example, pressing the F1 key usually displays Help in IBM programs. Another technique is using icons so that readers do not have to remember codes or program names. For example, the trash can means "delete" to people using the Macintosh computer.

Aesthetics

Readers' willingness to use a document (printed or online) based on its appearance.

When all else is equal in printed materials, their physical attractiveness could tip the balance for readers. The resulting performance that communicators want readers to achieve with documents is unlikely to happen unless readers are motivated to use them. Attractive bindings, high quality paper, balanced page layout, interesting graphics, and legible print invite readers to use documents. These devices make documents look easy to use. Attractive documents look as if the producers value the information enough to care about its appearance and suggest that the same care has gone into writing the text as was invested in its design.

Aesthetic value may also be achieved by establishing design standards. Repeated design features create a positive, unified image of a company or writer. According to JoAnn Hackos (1988), companies convey a sense of stability when all documents look as if they belong together.

In online documents, fewer aesthetic issues influence design choices, because screens have been limited primarily to black and white contrast with one style of type. As described in Chapter 8, color, sound, and motion are introducing the aesthetics of film and broadcast television to online documents, opening a new world for writers and designers.

The performance of readers in learning and completing tasks depends on their seeing, finding, understanding, remembering, and enjoying information. To successfully elicit performance, technical communicators can use an arsenal of visual elements to complement the verbal ones.

This chapter divides these design elements into two categories: "big-picture" techniques and "little-picture" tools. No matter how the elements are classified, in practice, visual features work in concert to help readers use documents. Technical communicators integrate both big-picture and little-picture design elements when creating document standards. *Standards* are an established set of physical features of the text that signal structure and set the tone for readers. Standards ensure consistency by addressing such issues as type style and placement and such patterns of visual cues as headings and icons. This visual consistency establishes expectations that help readers understand how parts of a document are related and how to locate information.

THE BIG PICTURE

> "As you begin to plan a document, you should not only consider who your audience is and how they will use the document, but also *how you can use the visual presentation of the text to reinforce the structure of the information.*" (italics added)
>
> (Benson, 1985)

Whether you are developing a written or online document, you use visual techniques to establish the big picture for an entire document. The big picture explains

- What the information is about
- How it is organized
- How readers can use it

The big picture sets readers' expectations—what using the document will be like, what to expect, and where to find information.

Big-Picture Techniques for Printed Documents

At the big-picture level you can signal the structure of a document by combining techniques of *chunking, layering,* and *outlining*. Each of these suggests a different visual technique for handling information.

Chunking Breaks up Content. A chunk is a segment—or logical unit—of information. It can be any size—a section of a book, a group of paragraphs and illustrations under a heading, a paragraph on a page, a computer screen of information. Chunking divides information into manageable units for readers. For example, simple paragraph divisions often help readers break text into meaningful units. Lists create clear chunks, too, as do tables and boxed information.

By using these and other chunking techniques, you help readers segment and relate pieces of information. You also help readers find information. These design elements can simplify retention, comprehension, and accessibility.

Layering Signals Relationships. A layer is a level of information in a hierarchy. Layering information signals order and importance. Descriptions of document design include techniques for layering information: *queuing, filtering, mixing modes,* and *abstracting* (Keyes, Sykes, & Lewis, 1988). Chapter 1 describes these concepts in terms of understanding readers; we are reviewing these concepts in terms of design.

- In *queuing,* you arrange chunks into a spatial hierarchy by using graphic elements such as headings and rules, or lines, to stress the order and importance of the material for the reader.
- With *filtering,* you further segment material by adding more visual layers through such graphic elements as subheadings, lists, color, and type changes.
- Through *mixed modes* you translate the filtered information to appeal to different cognitive (or thinking) styles by incorporating a range of visual techniques such as charts, diagrams, illustrations, formulas and the like.
- By *abstracting,* you show relationships among the smaller units.

Techniques for chunking and layering include:

Tabs. *Tabs* group text into physical units and help guide readers through lengthy text. They may appear as separate pages, or divider tabs, of heavier paper stock, often with a stub protruding from the edge. They may simply be marked as heavy printed bars on the outer page edges, as is illustrated in the outer right margin in Figure 7.1.

This example from a procedures manual shown in Figure 7.1 uses tabs to group tasks by job title. Employees use these tabs to find the procedures assigned to their jobs.

■ *Process Cash Receipts*

Adding a Cash Receipt Transaction

Summary

Once you create a packet, you can add, modify, scroll, print, or delete information up until the time you process the packet. When you process a packet, you add the information in it to the general ledger.

To add a cash receipt document you must complete the information on two screens, the document header screen and the distribution information screen.

51 Packet Handler

Step 1: Open a cash receipts packet.

Enter the ID for the current cash receipts packet. If a CR-type packet does not already exist, create one, page 29.

Step 2: Begin entering transaction information.

Enter a valid fund number for this document, a unique document number, the date of the cash receipt, and a description of the transaction.

Step 3: Enter distribution information.

Enter the general ledger function number for the first distribution. Enter information in other fields as appropriate.

Step 4: Balance the transaction.

Check the running totals of debits and credits at the bottom of the screen. If they are not the same, the document is out of balance and you must correct it before the system can accept it.

When you have entered all the distribution lines, press Enter at function. Check the information on the screen and indicate whether you want to accept, reject, or change the document.

Step 5: Add another transaction or go to another procedure.

Step 1 packet information

```
********* REVISION  4.33 ********* L.G D.P.C. ************ 07/25/89 ************
                         Accounting System
                         Post Transactions

     Packet ID: MCM                          Trans Type: CR

     Enter Budget Percentage To Check Against :    25.00%

               This Will Be A New Packet

               Is This Screen Correct? Y/N: _
```

Step 2 transaction information

```
********* REVISION  4.33 ********* L.G.D.P.C. ************ 07/25/89 ************
                         Accounting System                 Add
                          CR Transactions

 Trans Type : CR    Fund : 101    Sub Fund :         Doc # : 058847

 Doc Date : 07/29/89                   Beginning Balance Code :

      Document Description : ____________________
```

Step 3 distribution information

```
********* REVISION  4.33 ********* L.G.D.P.C. ************ 07/25/89 ************
 Trans Type : CR                Fund : 101          Document : 098675
          Sub  Sub  Sub                                               Line
 Funct-Obj-Fnd-Funct-Obj           Debit Amt        Credit Amt        Item
 ____-  -  -  -                        0.00              0.00         001

              ------Previous Distribution Line------
          Sub  Sub  Sub                                               Line
 Funct-Obj-Fnd-Funct-Obj           Debit Amt        Credit Amt        Item
     -  -  -  -

  Total Debits -              Total Credits -
```

Process Cash Receipts

page 4 LGDPC

General Ledger Processing page 5

FIGURE 7.1 Two-page Spread (or Folio) from a Procedure Manual Illustrating Chunking, Layering, Page Layout, and Type (Courtesy of Local Government Data Processing Corp.)

Headings. *Headings* are words set apart from body text to identify topics. Headings entice readers to use a text by making it easy to dip in, skip around, and backtrack—the way most readers use a text. Headings serve two functions, one informative and one structural. They label chunks and alert readers to a topic change. They are often set in boldface type to capture attention, and they serve as signposts to set up the structure or framework of a document.

Levels of headings show layers or ranking of information. Consistent placement and type style of headings set up patterns on a page or screen to let readers know where to anticipate information. Headings help readers choose which section to read and help them remember the structure of a text.

Research confirms that readers, especially readers who are in a hurry or who have a low skill level, benefit from headings that are full statements or questions (Benson, 1985).

Design manuals cite a variety of functions and characteristics of headings. In *Graphic Design for the Electronics Age*, White (1988) describes thirteen different types of headings. The following describes three common types.

Type	Description	Example
Structural headings identify the largest chunks of information.	Running headers and footers appear in small type at the top or bottom of pages or screens and repeat information such as topic, page number, and date. A typical word processing screen includes a line to identify the name of the document and where the user is currently working in that document. Standing heads are a separate line above the text, usually set off by a different typeface or other graphical feature. Periodicals, such as magazines and newspapers, use standing heads to identify regular departments such as editorials and advice columns.	Figure 7.1 illustrates using a header to identify the chapter ("Process Cash Receipts") and a footer to cite pages, identify the company ("LGDPC"), and name the system being documented ("General Ledger Processing").

(*continued*)

Layered headings establish hierarchies in text.	Changes in typeface, size and placement signal different heading layers. Main headings, also called first-level headings, indicate the largest divisions in a text. Main headings are traditionally capitalized and centered. Subheadings are lower-ranking headings and set off subsets of the main topic. Second-level headings often appear on a separate line, and third-level ones are cut into the text. These choices were dictated by typewriter technology. Desktop publishing expands options for size and style of type, allowing you to present more levels of headings with more variation. This makes hierarchies more readily apparent.	Formal scientific reports, for example, use three to four layers to move from main chunks, such as Methodology, Findings, and Conclusions, to subordinate levels that segment these parts. In Figure 7.1, for example, four different levels of headings are indicated through changes in size and style of type and placement of text.
Emphasis headings highlight information and clearly separate chunks.	Hanging heads appear in the left or right margins of asymmetrical layouts, often in newsletters or procedure manuals. Placing headings in margins helps readers skim and skip text. Hanging heads work well with shorter, more readable line lengths. Over-heads, a line of type above the title and usually separated by a rule, or line, create emphasis by identifying the topic of an article or using a catchy phrase to attract attention.	The steps in Figure 7.1 "hang" in the margins.

Lists. A *list* is a chunk of information clearly separated from the rest of the text by form and substance. Lists increase comprehension and accessibility. A list, for example, might indicate the relationship among steps in a process, summarize recommendations, or describe product benefits. Lists are especially helpful in online documents because readers scan them quickly, and scanning is the predominant method of reading online (Horton, 1990).

By grouping related items in a list, readers can more easily see that all included items are similar

- Conceptually.
- Grammatically.
- Typographically.

List items may be introduced by bullets, numbers, letters, or words. Depending on the purpose, list items may contain only single elements such as names, dates, or locations, or they may include short paragraphs such as some of the lists used in this chapter.

By following a few guidelines for the layout of lists, you can make them easier to read:

- Indent the list to align the text so that the numbers or bullets stand out in the left margin.
- Avoid setting lists in type smaller than the body text unless you want the reader to consider the information less important than the rest of the text.
- Punctuate consistently. There are several acceptable styles, so pick one and use it consistently, or follow the guidelines in your company's style manual.
- Use the following cues before list items (Carliner, 1987):

    ☐ Checkoff box, if each item must be performed

    ■ Bullet, if list items are equally important

    1. Number, if list items should be considered in a certain order

Outlining Techniques Make the Document Structure Visible

When you reveal the outline of a text you show readers its structure. Knowing how a document or program is structured helps readers understand and find information. It also helps them recognize and remember hierarchy and relationships. Usability tests suggest that readers rely heavily on outlining techniques such as tables of contents, indexes, and tabs (Ramey, 1988).

Specific techniques for outlining include:

Table of Contents. Previews the structure of a document so that readers can determine whether the document meets their needs.

Example: Software Tutorial Manual for New Users—provides a preview of the lessons

Index. Lists subjects to help readers locate information within the text. Usually found at the end of a document or by pressing a function key in an online document.

> **Example:** In a software reference manual organized by menus, an index helps readers find a specific menu.

Page Separator. Divides one section from the next with a blank page, often of a different color or weight. Gives readers a rest and a "pat on the back" for completing a section.

> **Example:** In training materials, page separators provide a physical pause that encourages the reader to get up and take a break.

Glossary. Defines information presented in the main text, but not necessarily explained there. Provides a more detailed level of selected information such as definitions, field names, and codes, without cluttering the main text. Chapter 5 explains how to write glossaries.

> **Example:** Textbooks usually have a glossary.

Appendix. Provides supplemental information without cluttering the main text.

> **Examples:** Business proposals typically attach a list of clients and completed projects. Research reports might append charts of raw data and questions.

Big-Picture Techniques for Online Documents

To access, comprehend, and remember online information, readers need to understand the parts of the system and how they relate (chunks and layers) as well as how the overall system is organized (outlining). To help readers achieve these results, technical communicators need to adapt some design approaches used in print and invent new ones, especially for interlinked hypertext documents (Bernstein, 1991). Specific design features create chunks, layers, and outlines at the big-picture or "macro systems" level.

Chunking and Layering in Online Documents Segment and Link Information. Techniques to create chunks and layers of online information borrow from print media and introduce unique design solutions. These techniques must control the amount of information presented at one time.

Menus. Pull-down screens or windows allow users to remain in a document while selecting another feature, task, or information source, such as a help screen or encyclopedia reference.

Headings. Text headings in the margins (as on paper) label chunks of information, such as steps in a process, but headings also set up frames or templates on a screen that remain consistent from screen to screen to support data entry or facilitate access. For example, to help a user set up a new account online, each screen should use the same style and placement of headings to identify the needed information such as name, account number, and address. Graphic cues complement or even replace headings in many instances.

Outlining Techniques Online Create Patterns for Navigation. Because readers cannot see the complete document as a whole and flip through pages to know where they are in the text, they need to have a mental picture of the whole system and understand the ways to interact with the system ("user interfaces"). We call the following techniques outlining because, like an outline for a printed document, they signal structure and keep the reader on track.

Metaphors. Metaphors convey a mental model of the structure. Because readers cannot literally see and touch online documents, they must interact with a concept of what happens "in the electricity." The perception of the structure of the system is called "virtuality" (Horton, 1990).

Readers come to online documents already comfortable with reading books. You can capitalize on this experience by incorporating familiar indexes, illustrations, and typographic cues. You can also introduce other visual techniques to create structure for online information. Hypertext, for example, uses the image of cards, stacks of cards, and networks of stacked cards. The desktop metaphor for the Macintosh uses items related to an office, such as folder, calendar, and dictionary (Horton, 1990).

Repeated cues. Each screen repeats information, as running headers and footers do in paper documents. This repeated information is a memory anchor. These anchors tell readers how to move forward and backward and identify the location in the online document. Place repeated cues consistently on a screen. For example, always place action on the same line within a program. Graphic elements, such as icons and bold type, should separate repeated cues from content text.

THE LITTLE PICTURE

Visual tools not only help readers understand and use entire documents, but they also assist readers at the page or screen level. These tools help readers find information and read it more easily.

At this level you can use the same visual tools you use at the "big picture," or document, level to help readers. The individual building blocks—the tools for presenting the visual picture—are placement, graphic elements, type, illustrations, and color. Chapter 6 covers how to use illustration and color. We cover placement and white space, graphic elements, and typography.

Technical communicators can learn from the graphic designer's approach to achieve visual balance and rhythm. In an ideal design, readers can squint at the page or screen and see a clear visual structure. The next sections of this chapter describe the tools for creating effective pages and screens.

Little-Picture Tools for Pages

At the page level, you have a variety of tools to influence how readers interact with the text. Placement, graphic elements, and typography work together to create unique designs to help readers achieve results.

Placement Guides Readers. Placement is the location of text, illustrations, white space, and other graphic elements on the page or screen. Placement decisions are influenced by general theories of how readers read, the purpose of the document, and the aesthetics of proportion, balance, rhythm, contrast, and unity.

Depending on their culture, people learn to read in specific patterns. Readers of English not only read from left to right, but they also scan a page in a Z pattern and tend to see the center of a page (the "optical center") as slightly above the absolute center.

Knowing where the eye tends to go can influence where you place text, visual heads, and other elements of the page. Placement must factor in different needs to achieve comprehension, accessibility, and retention. Whether the design problem relates to an annual report, a financial newsletter, or VCR instructions, the placement tools remain the same: the grid, white space, and margins.

Tool	Description
Grid layout is a series of horizontal and vertical lines that sets the pattern for columns, headlines, margins, copy, and illustrations.	The grid creates a blueprint for layout that can be repeated page after page for consistency. This consistency helps readers find and remember information. Grids are usually flexible enough to allow some variety so pages look similar but not cloned. Page design is not limited to the standard 8 1/2" by 11" page. You can design a two-page spread, called a *folio*, as a single unit. A folio is illustrated in Figure 7.1.

(*continued*)

<table>
<tr><th>Tool</th><th>Description</th></tr>
<tr><td>White space is the blank area on the page or screen.</td><td>White space is carefully planned empty areas on a page. White space

■ Guides readers through the structure. Deep top margins, for example, signal beginnings of new sections and chapters; indented lists indicate subordinate but important material.
■ Prevents fatigue. Readers tire from forging through the unbroken gray space of text. They may even refuse to read at all. White space provides rest stops that make reading a more pleasant task.
■ Creates aesthetic appeal. Handbooks on graphic design explain how equal portions of blank areas create balance, repeated patterns achieve rhythm, and shifts create contrast.

White space is created by margins, indentations, and space between columns and paragraphs, around headings, and between lines of type, as shown in Figure 7.2. Experts recommend filling 50% of a page with white space.</td></tr>
<tr><td>Margins, a special type of white space, are the unprinted border on the outside edges of the text or page.</td><td>To create interest and break monotony, margins for a page should be uneven, with the bottom or foot margin usually the largest. Once you set outer margins, they should remain constant throughout the document.

A justified margin means that all lines begin or end at the same place.

A ragged margin is visually "ragged" or uneven. The rule of thumb for standard text is to set the left margin justified (that is, the left side of the column is justified); and the right margin ragged. Having an even line on the left edge gives the eye a clear place to begin. Although the question of ragged or justified right margins is debated, ragged right margins have two advantages:

■ Reading ease. People read by recognizing groups of letters, so it follows that reading ease can be reduced if natural spacings are interrupted. Justified text forces unnatural spacing to make the lines even.
■ Aesthetics. Unnatural spacing causes white "rivers" to run through text. These occur when the word processor adds spaces to force lines to even out. Adding a hyphenation option can reduce the maximum raggedness caused by some word lengths.</td></tr>
</table>

Graphic Elements Break Up Text. *Graphic elements* are lines, borders, circles, and other shapes without illustrative content that are used to organize a page. Graphic elements emphasize, isolate, and separate information to aid learning and retention. For example, tables and

EXCEPTION: "Doing Business As" (DBA) accounts use the Social Security Number of the individual.

(Refer to the *Platform Manual* for detail on Debtor-in-Possession Accounts.)

Interest on Lawyers Trust Accounts (IOLTA)

An IOLTA account holds interest earned on certain accounts held by lawyers. The funds are used to provide legal services for the poor and for programs to improve the student administration of justice for law loans and scholarships.

The Branch Manager is responsible for ensuring Branch personnel

- Use "Tennessee Bar Foundation" in the legal title for IOLTA accounts, in the following format:

  Tennessee Bar Foundation
  IOLTA Account for
  (Title of Law Firm in the legal title).

- Use Tax ID number, 62-6074501 (Tennessee Bar Foundation).

A W-9 is not required for an IOLTA account because the Tennessee Bar Foundation has a W-9 in central file.

(Refer to the *Platform Manual* for further detail on IOLTA Accounts.)

OTHER MATTERS AFFECTING DEPOSIT ACCOUNTS

Reporting on Deposits Held by Foreigners

The Federal Government requires First American to report on accounts held for non-resident aliens. The Branch Manager is responsible for ensuring Branch personnel

- Obtain proper identification (a green card) for non-resident aliens.
- Obtain a W-8 form for each holder listed in the legal title for accounts opened for non-resident aliens.
- Use Federal Withholding Code 7, "non-resident."

If an account has a foreign and domestic owner, Branch personnel must list the domestic owner on the first line of the legal title and use the Social Security Number of that owner.

FIGURE 7.2 Procedure Manual Page Illustrating Layout—White Space (Courtesy of First American National Bank, Nashville, TN)

matrices presented with intersecting lines highlight information and separate it from text.

Graphic elements can signal the structure of text or relationships within texts. For example, changing the width of a *rule* (a special type of line) can indicate a change in topic.

OPERATION

IMPORTANT
Review and understand the warnings in the Safety Information Section. They are needed to safely operate this heater. Follow all local codes when using this heater.

To Start Heater

1. Follow all installation, ventilation, and safety information.
2. Locate heater on stable and level surface. Make sure strong drafts do not blow on heater.

⚠WARNING

Fully close main burner valve before lighting pilot. If not, severe burns can occur. Do this by turning valve handle towards the OFF position until it stops.

3. Open propane supply valve on propane tank(s) slowly. *Note:* If not opened slowly, excess-flow check valve on propane tank will stop gas flow. If this happens, close propane supply valve and open again slowly.
4. Push in and hold safety control valve button. Push piezo ignitor button. The piezo ignitor button may need to be pushed 3-8 times until the pilot lights. *Note:* Hose may be filled with air. If so, keep safety control valve button pressed and wait 20 seconds before pressing ignitor button again.
5. When pilot lights, keep safety control valve button pushed in. Release button after 30 seconds.

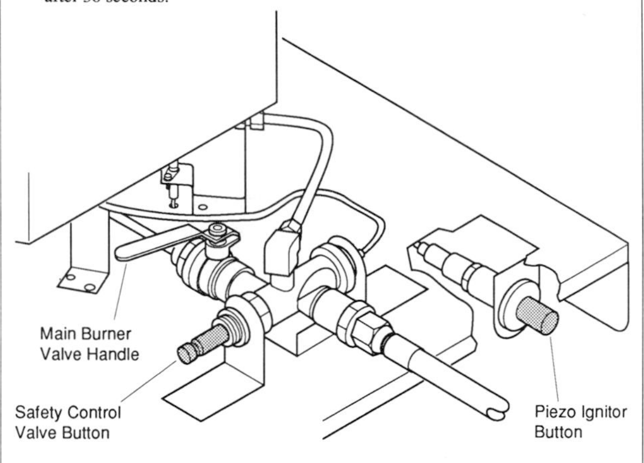

Figure 4 - Safety Control Valve Button and Piezo Ignitor Button Locations

Continued

7

FIGURE 7.3 Equipment Manual Page Illustrating Graphic Elements and Type (Courtesy of DESA International)

Graphic elements, such as borders around illustrations and boxed-in text inserts, as shown in Figure 7.3, highlight ways to achieve aesthetically balanced designs.

The following table describes some key graphic elements and how to use them.

Tool	Description
Rules separate information.	Rules are lines that separate text. These design tools delineate chunks, define white space, and organize and emphasize information on a page. Rules can run horizontally or vertically and appear in varied thicknesses. Use rules ■ Above a heading to separate sections of text. ■ At the bottom of a page to separate the page number and footer from the text, as in Figure 7.1. ■ To set off "pull quotes," short summaries of key points embedded in a text. For example: "Ask not what your country can do for you . . ." ■ To separate information blocks as in Information Mapping™ (Horn, 1976). ■ To separate columns of text in newspapers and magazines.
Borders and circles create self-contained units that isolate a part of the text for emphasis or contrast.	Borders can be three-dimensional, shaded, squared, angled, or rounded. When considering a border, ask whether the text needs isolation. In newsletters, for example, borders often ■ Separate publishing credits from the text. ■ Separate short, self-contained articles and illustrations from the main text. ■ Emphasize special announcements. Readers can benefit when warnings and illustrations are isolated. The operator and safety instructions in Figure 7.3 illustrate a page border, boxes and graphic elements to draw the reader's attention to critical information.

Typography Creates Both Clarity and Interest. *Typography* refers to the look and legibility of the letters used, including type style, shape, size and the placement of type on the page. These elements work together to create a legible, attractive text to help readers find information, comprehend it, and read without tiring their eyes. An appropriate type choice can even motivate readers to read in the first place. Desktop publishing, in particular, provides a variety of options for using type.

The following table describes some key tools of typography and how to use them.

Tool	Description
Case includes lowercase and uppercase characters.	Lower case letters have *ascenders* ("b" "d") and *descenders* ("p" "g") that create a distinctive shape for a word. The irregular outline of the shape aids the reader in recognizing the word and can speed up reading.

(*continued*)

Tool	Description
	Uppercase letters signal the beginning of a sentence. They can also highlight a short phrase and are used in trademarks, logos, and labels. Most text is set in a combination of upper- and lowercase letters for two reasons: ■ All uppercase text looks intimidating because it lacks the distinctive outline of mixed upper- and lowercase letters. When all words have a box or rectangular shape, the reader lacks cues to help identify words. ■ Uppercase letters use space less efficiently because they are bigger and wider than lowercase letters. Also, extra space must be added between lines of all capital letters to increase legibility (White, 1988).
Typeface includes serif and sans serif characters.	*Serifs* are the extensions or little "feet" attached to ends of lines in letters. More than just decorations, the extensions group letters and pull the eye along the page. For example, Times Roman is a serif typeface: This is Times Roman. *Sans serif* type, as the word *sans* (without) suggests, lacks the decorative extensions. The clean, bold appearance is often associated with contemporary style. Helvetica is a sans serif typeface: This is Helvetica. Graphic design manuals often suggest using serif type for blocks of text and sans serif for headlines. The rationale is that serif type is easier to read in large blocks of text because the extensions link letters for readers. But the final word is by no means set for this choice. Ease of reading and comprehension are affected by a long list of other features, including type size, line length, paper finish, ink, and so on. So what should you consider when choosing serif or sans serif type? In *Graphic Design for the Electronic Age*, White (1988) sifts through the pros and cons and suggests: ■ Base your decision on what is familiar to your readers. Consider that documents in the United States most often use serif typefaces and those in Europe use sans serif type. ■ Use a type that conveys an appropriate tone or style for your purpose. Although sans serif type is difficult to read in long segments, it can be used to advantage in short paragraphs. The clean lines of sans serif, for example, work well for the procedure manual shown in Figure 7.1. The unfamiliar type style suggests that this manual "isn't like all textbooks," a comforting message to novice learners.

(*continued*)

- Consider how much information must fit on a page. The horizontal extensions of serif type allow less space between lines without losing legibility. More space is needed between lines to achieve equal legibility with sans serif type.
- Consider the need for variety. If you need subtle variations and a variety of designs, you will find more choice among serif type faces.

Font size is determined by the height and width of characters in a typeface.

Font size refers to the size of type. *Point* is the unit for measuring font size. A point is equivalent to .014 inches. Graphic artists use the terms font size, type size, and point size interchangeably. Font size affects legibility. Use these guidelines for choosing the most legible font size:

- For headlines, titles, and the like, use a larger font size than appears in the text.
- For text, standard type size ranges from 9 to 12 point.
- Consider the impact of the *x-height* of letters on body text. The x-height refers to the main body of the letter between the ascenders and descenders.

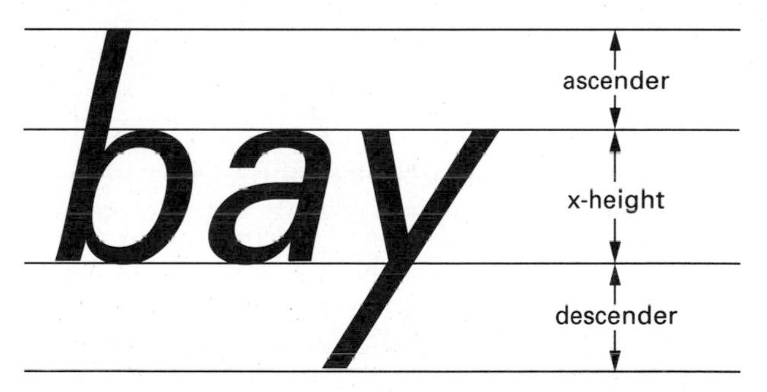

  The higher the x-height, the larger a font appears. The advantage to using a font with a larger x-height is that you can reduce the size of the font from say 12 point to 10 point to get more lines on a page without sacrificing legibility.
- Remember that column width also influences the choice of font size. As a general rule, the narrower the column, the smaller the font size you should use.

Font weight is determined by the thickness of the characters in a typeface.

Bold, outline, and shadow are ways to vary the weight of a font. Varying font weight signals a change in information, such as headings, titles, and illustrations.

Bold fonts are more effective for highlighting than using all uppercase letters or setting lengthy sections in italics. Figures 7.1 and 7.2 use bold headings to emphasize changes in topics.

Spacing is the distance between characters and between lines of type.

Spacing influences reading ease by determining whether text appears cramped or overly expansive, both of which increase the difficulty of reading.

Typography has two terms to describe spacing:

- *Leading* (pronounced lĕd-ing) refers to adjusting the space between lines. The shorter the lines, the less

(*continued*)

Tool	Description
	space is needed between them. Serif type requires less leading than sans serif type because the extensions pull the eye along. But if the x-height is tall, the letters take up more space on the line, so more space may have to be added between lines. Leading is usually two points larger than the type size and is indicated by the expression typesize/leading. For example, if you use 10-point type, you would probably use 12-point leading, indicated as: 10/12. Headlines or headings that cover more than one line present a different leading problem. The automatic spacing on most desktop publishing systems leaves too much space between lines for the headlines to look like a unit. If you have the computer capability, you might want to close the space between lines to improve readability. ■ *Kerning* refers to adjusting the space between letters. Some letter combinations create extra space and are therefore hard to read as a unit, especially in headlines.
Line length is the width of a column of text.	The decision about what makes line length comfortable for readers depends in part on other type features: typeface, size, and leading. What readers expect and how they will use the document also affect choices. For most printed documents, research suggests that 50 to 70 characters per line is best. Long line length makes it hard for the eye to stay on track and return to the beginning of the next line without having to reread.

Little-Picture Visual Tools for Screen Design

Layout and typography are as important to online documents as they are to printed ones, but limitations of the computer screen change the applications. Combining characteristics of text and film, online documents are constrained by the size of the screen and resolution quality, but offer new capabilities of sound and motion (Rubens & Krull, 1985).

The following chart presents considerations for designing screens:

Tools	Description
Placement tools include grid layout, white (blank) space, and margins	A screen grid is 80 characters by 24 lines. Given the guideline of covering only 50% of the surface with text (and some research suggests using even lower densities online), the amount of space for producing readable screen text is very limited.

(*continued*)

Few screens offer variations in style and size of type or the ability to adjust space between lines. As the technology advances, so will the design choices. The following guidelines reflect current technology (Horton, 1990; Rubens & Krull, 1985).

- Limit line length to 60 or fewer characters.
- Use less than 50% of the available space on the screen for type.
- Create white space by putting blank lines around text and freeing the left area for cues.
- Keep paragraphs short (three to six lines), indent them, and separate them with a blank line.
- Use the left margin for headings and other cues.
- Balance screen displays symmetrically.

Graphic elements and typography

Use graphic elements and typography to speed up reading, prevent fatigue, and improve aesthetic quality on screen. Design choices include the following:

- Use upper- and lowercase when available.
- Use boldface or color to highlight headings and other cues.
- Use flashing characters sparingly.
- Use windows and other graphic aids to group related material.
- Highlight information by introducing variations such as flashing messages, reverse video, and pull-down menus.
- Use color, *but*
  - Avoid complementary combinations (such as blue and green).
  - Limit choices to four or five colors.

CASE STUDIES OF EFFECTIVE APPLICATIONS OF VISUAL TOOLS

The following cases illustrate ways to integrate design techniques and tools as building blocks to solve practical problems. For each case, test your understanding of visual techniques and tools. Review the situation, then sketch out how you would design the technical communication, in print or online. Use the questions to help you consider all the angles.

Case 1: Sales Brochure For a Technical Product

Situation. Your company sells telecommunication networks for assisting organizations in sharing data to make complex business decisions. Your boss, the vice-president for product development, asks you to create a product brochure to be handed out at the annual trade show. Your boss is worried that the advertising writers your company usually uses will not be able to explain the technical aspects of the system. How do you start?

Audience Analysis. First, you review audience characteristics. You know that telecommunication managers and technology directors attend the trade show. They tend to be very knowledgeable and interested in product features, product test histories, and prices. Besides working with technology all the time, they are sophisticated computer users as well. Even though your boss asked for a brochure, you decide to use a multimedia approach because

- The audience will respond positively
- Nonsequential access to information will better suit your audience, some of whom will want to examine a very detailed test history and others who will not want the detail

Strategies. There are many considerations to designing online documents, but, for this exercise, just consider the visual aspects of your screen design. As you plan your screen layouts, answer the following:

- What techniques will you use to link screens and types of information?
- How will you summarize data—graphs? paragraphs? other ways?
- How will you use color?
- How will you integrate text and graphics on screen? How will that look? What graphics will you use? How will they interact with the sound track?
- How much data will you display on one screen?
- How will you emphasize text on screen?
- How will you vary the presentation?

One Solution. Faced with a similar problem, Bob Duthie of Duthie & Associates in Nashville, TN, created a multimedia presentation for South Central Bell that allows viewers to examine the impact of telecommunications in a variety of industrial settings. The viewer can examine summary or detailed information. See Figures 7.4 through 7.6.

Case 2: Software Reference Materials

Situation. You work as a technical communicator for a state government welfare department. Employees of your department are located all across the state and take applications for federal and state welfare programs. Your department was recently ordered to become a paperless operation. Under the new program, employees will no longer complete paper forms. Instead, they will enter the application information

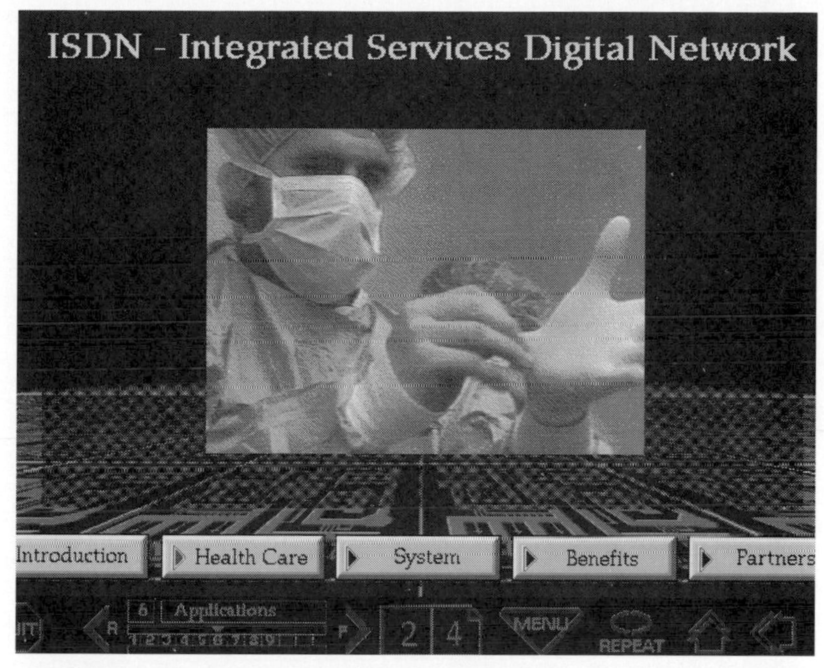

FIGURE 7.4 Case 1—Sample Screen Introducing Healthcare Application (Courtesy of South Central Bell Telephone Co.)

FIGURE 7.5 Case 1—Sample Screen Illustrating X-ray Image Shared Through Telecommunications (Courtesy of South Central Bell Telephone Co.)

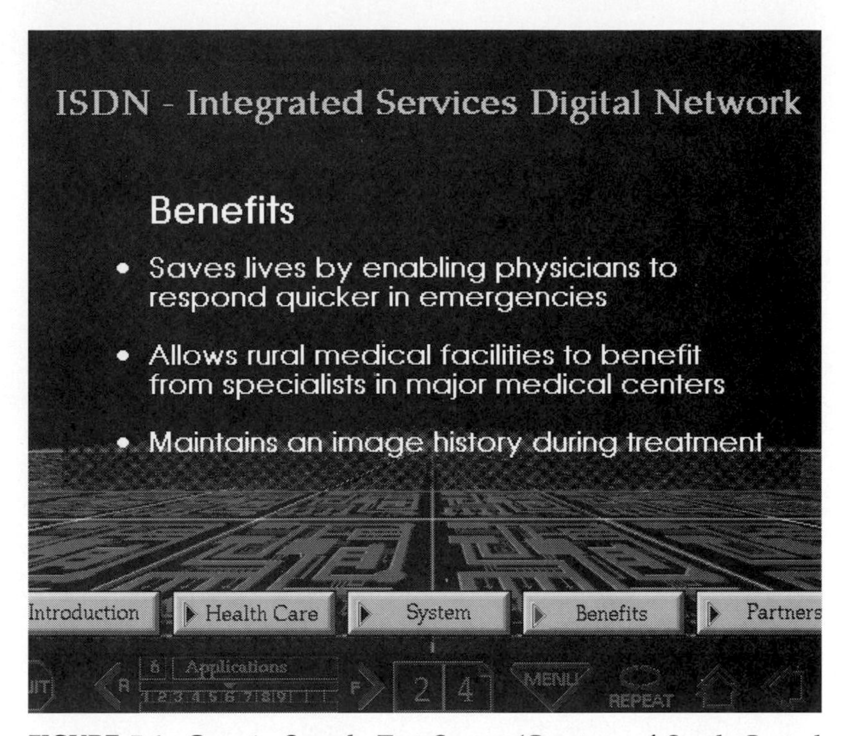

FIGURE 7.6 Case 1—Sample Text Screen (Courtesy of South Central Bell Telephone Co.)

directly into terminals that transmit the data to a centralized mainframe system. Your boss, director of technical support, wants you to create a 100- to 200-page reference manual that will be the employees' only user's guide for the system. How do you begin?

Audience Analysis. You know your audience pretty well. While most will be in support of this change, many of the field personnel have never used a computer terminal. Also, the department has decided that it will provide some initial system training, but most employees will have only your reference guide as a resource. The software is complex and not always intuitive. Some employees are pushing for online help in addition to the 100-page reference guide.

Strategies. As you design the organization and layout of the guide, you also consider how the help screens will look and how the two types of communication will relate to one another.

- What information belongs in the manual and what belongs on screen?
- What visual techniques will you use in the manual and online?
- What standard information will appear on each page, online, and in both places?
- How will you emphasize information on the page and online?
- Will you use graphic elements? What are your screen constraints?

One Solution. The Tennessee Department of Human Services developed the following reference materials when faced with a similar situation and constraints. They created a reference guide, shown in Figure 7.7, that presented the structure of the system, and they created online help screens, shown in Figure 7.8, that described items on the screen.

ACCENT User Manual

How to Get Help
ACCENT provides two important sources of help for you if you don't understand what is displayed on your screen or what you are supposed to type in response to a prompt. Of course, if you have questions about general subjects which are not specific to any one screen, then this manual should be your first source of information.

Otherwise, ACCENT has what we term 'screen help' and 'field help.'

Screen Help

What it is: One or more screens of help text pertaining to the ACCENT screen you are viewing at the time of your request for help

What it does: Displays help information for a given screen in the following format:

1. Purpose — One or more sentences describing the intent of the screen
2. Citations — A list of policy citations which pertain to the content of this screen
3. Field Definitions — For each field on the screen, a description of the field and what is expected by way of data entry (Also see 'Field Help' below.)

<Tran Code> <Screen Name>

Purpose: ______
Citations: ______
Field Definition
___ ______
______
___ ______
______
<More>

How to get it: At any time, press the PF1 key on your keyboard. (Look at the terminal keyboard layout in this guide if in doubt about the PF1 key). While in the help facility, you can use the following keys:

PF8 To advance forward to the next page of help information in the sequence, if available

PF1 To *return* to the data entry screen from which you requested help, i.e., to exit help

Field Help
A field is a single data entry area on the screen. For some fields, the information which you are expected to supply is coded. For example, verifications require a two character code. Codes are pre-defined, meaning that you must use one that ACCENT recognizes.

What it is: A list of the valid codes which you can use in the field in question

What it does: Displays one or more full screens of valid codes for the field in question with a description of the meaning of each code.

How to Get Help 1/19/90

FIGURE 7.7 Case 2—Sample Page from User Guide (Courtesy of Tennessee Department of Human Services)

```
SFHM                              HELP SCREEN FOR AEIID
                                       Page 4 of 5

Fields
Continued: "S"

           Individual's sex;  M = male  F = female

           "R"

           1  character code (from TETC table) for individual's race; press #
           when on field for list of codes

           "LIV ARR TYPE"

           2 character code (from TLAR table) for individual's living
           arrangement; press # when on field for list of codes



Next Tran:          Parms:                                          More:
```

FIGURE 7.8 Case 2—Sample Online Help Screen Complementing the User Guide (Courtesy of Tennessee Department of Human Services)

Case 3: Industrial Training Materials

Situation. You are a technical writing consultant. Your client manufactures a part for an automobile company with strict quality standards. The client has recently purchased a new piece of equipment for the assembly line. You have been retained to create materials that line workers can use as a job aid (a quick reference on the job).

Audience Analysis. The line workers' experience with the new equipment varies. Some have worked with a similar apparatus; others haven't. The reading-level of the workers ranges from 4th grade to 12th grade.

Strategies. You decide to create a quick reference booklet that can be laminated and displayed in the work area in easy reach of the workers. To do this you need to consider the following:

- How will you organize the information? By segments of the manufacturing process, by parts of the machine, or by quality guidelines?
- What visual elements will you use? For example, will you use color?
- How will you lay out the pages? What size should the pages be?
- What fonts and type sizes will you use?

One Solution. The training department at the Ford Glass Plant in Nashville, TN, created the quick-reference booklet shown in Figure 7.9. It shows effective use of illustration, type, layout, and organizational elements.

We have shown you only *one* of the many effective design solutions to each case study. One way to narrow your choices is to present a

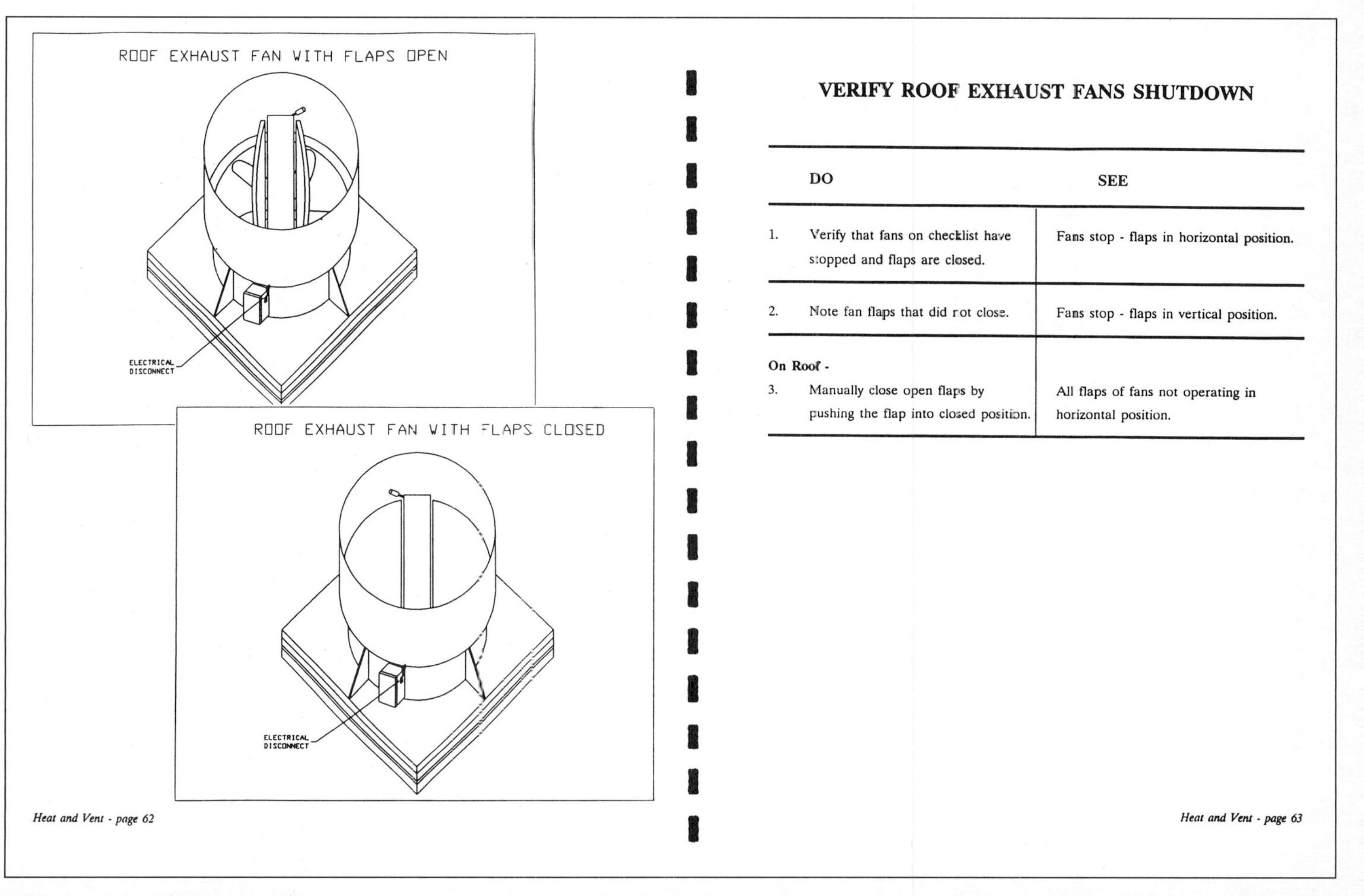

VERIFY ROOF EXHAUST FANS SHUTDOWN

DO	SEE
1. Verify that fans on checklist have stopped and flaps are closed.	Fans stop - flaps in horizontal position.
2. Note fan flaps that did not close.	Fans stop - flaps in vertical position.
On Roof - 3. Manually close open flaps by pushing the flap into closed position.	All flaps of fans not operating in horizontal position.

Heat and Vent - page 63

FIGURE 7.9 Case 3—Sample Industrial Training Procedure from Quick Reference Booklet (Courtesy of Ford Motor Company)

proposed solution to a small group in a structured interview, called a "focus group." Another method is to perform usability testing, described in Chapter 11.

SUMMARY

This chapter explained how effective design benefits readers and described specific design techniques and tools.

Effective design achieves these results with readers:

- Reading ease—readers' ability to see and read information
- Accessibility—readers' ability to find a specific piece of information in a document
- Comprehension–readers' ability to understand information
- Retention—readers' ability to recall both information and its location in a document
- Aesthetics—readers' willingness to pick up a document or venture into a program based on its appearance

Techniques that help readers grasp the big picture—structure and purpose—of a document include

- Chunking and layering—dividing information into manageable units through tabs, headings, lists, and menus
- Outlining—repeating the outline of the text through a table of contents, index, page separators, glossary, and appendixes.

Tools that help readers master the little picture—or read pages and screens more effectively—include

- Placement—effectively locating text, illustrations, white space, and other graphic elements
- Graphic elements—borders, circles, and other shapes without illustrative content that are used to organize a page or screen
- Typography—the look and legibility of letters used, including type font, shape, size, and placement on the page

LIST OF TERMS

abstracting A layering technique to show relationships among smaller units.

accessibility Readers' ability to find a specific piece of information in a document or program.

aesthetics Readers' willingness to pick up a document or venture into the program based on its appearance. Motivational aspects of document, including the enjoyment or pleasure readers experience from interacting with the document.

ascender The part of a character that rises above the body, such as the upper part of the letter *b*.

chunking A *chunk* of information is a logical unit of information grouped to help readers find information. It can be any size—a section of a book, a group of paragraphs and illustrations under a heading, a paragraph on a page, or a computer screen.

descender The part of a character that descends below the body, such as the lower part of the letter *y*.

filtering Segmenting material by adding more visual layers through tools such as subheadings, lists, color, and type changes.

folio A two-page spread designed as a single unit.

font The family of type, such as Helvetica and Times.

graphic elements Lines, boxes, circles, and other shapes that do not have illustrative content. Used to organize a page or screen.

grid layout An invisible grid that sets the pattern for columns, headlines, margins, copy, and illustrations. Helps the communicator create a blueprint for layout that can be repeated.

headings (headlines, heads) Tools to help draw readers into the text and make it easy to dip in, skip around, and backtrack. Headings signal to readers that the subject is changing. Headings also assist in scanning.

kerning Adjustments to the space between letters.

layering A technique of using levels of information in a hierarchy to signal order and importance.

leading Adjustments to the space between lines.

legibility Readers' ability to see and read information on a page or screen.

lists Special type of chunk. (Also see *chunking*.) Lists segment information into equal components and organize the components to reflect the meaning. Lists are clearly separated from the rest of the text.

margins Unprinted borders on the outside edges of the page.

mixed modes A tool for filtering information so that it appeals to different types of readers. Involves incorporating a range of visual techniques such as charts, diagrams, illustrations, and formulas.

outlining Makes the structure of the document visible to the reader; reminds readers of the structure, aids in understanding, and helps readers recognize and remember hierarchy and relationships.

point Unit of measurement for type. One point = .014 inches.

queuing Using tools such as headings and rules to stress the order and importance of the material.

retrieval aids Graphic cues used to search for information stored in a computer.

rule Line used as a graphical device, usually to divide parts of a page or screen.

sans serif Typeface that lacks decorative extensions at the ends of lines in letters.

serif Type face that has extensions or little "feet" attached to ends of lines in letters. Extensions group letters together and pull the eye along the page, making words easier to recognize and faster to read.

standards A set of physical features of the text that signal structure, set the tone, and establish consistency.

tabs Tools that provide quick access to sections of information by marking the borders of pages in those sections. Help to focus or limit readers' attention to specific information.

typography The look and legibility of the letters used, including type font, shape, size, and the placement of type on the page.

white space Blank areas on the page or screen. A design tool to guide readers, to prevent fatigue, and to heighten aesthetic appeal.

x-height The body of a character between the ascenders and descenders.

REFERENCES

Benson, P. (1985). Writing visually: Design considerations in technical publications. *Technical Communication, 32*(4), 35–39.

Benson, P. (1989). The expanding scope of document design. *Technical Communication, 36*(4), 352–354.

Bernstein, M. (1991). Deeply intermingled hypertext: The navigation problem reconsidered. *Technical Communication, 38*(1), 41–47.

Carliner, S. (1991). Lists: The ultimate tool for engineering writers. In D. Beer, (Ed.), *Writing & speaking in the technology professions: A practical guide,* (pp. 53–56). New York: IEEE Press.

Floreak, M. (1989). Designing for the real world: Using research to turn a "target audience" into real people. *Technical Communication, 36*(4), 373–386.

Hackos, J. (1988). Redefining corporate design standards for desktop publishing. *Technical Communication, 35*(4), 288–291.

Horn, R. (1976). *How to write information mapping*. Lexington, MA: Information Resources.

Horton, W. (1990). *Designing and writing online documentation: Help files to hypertext*. New York: John Wiley & Sons.

Keyes, E., Sykes, D. & Lewis, E. (1988). Technology + design + research = Information design. In E. Barrett, (Ed.), *Text, conText, and hyperText: Writing with and for the computer* (pp. 251–264). Cambridge, MA: MIT Press.

Miller, G. (1956). The magical number seven, plus or minus two: Some limits on our capacity for processing information. *The Psychological Review, 63,* 81–87.

Ramey, J. (1988). How people use computer documentation: Implications for book design. In S. Doheny-Farina, (Ed.), *Effective documentation: What we have learned from research* (pp. 143–158). Cambridge, MA: MIT Press.

Rubens, P., & Krull, R. (1985). Applications of research on document design to online displays. *Technical Communication,* 37(4), 29–34.

Rubens, P., & Rubens, B. (1988). Usability and format design. In S. Doheny-Farina, (Ed.), *Effective documentation: What we have learned from research* (pp. 213–233). Cambridge, MA: MIT Press.

Schriver, K. (1989). Document design from 1980 to 1989: Challenges that remain. *Technical Communication. 36*(4), 316–329.

Shirk, H. (1988). Technical writers as computer scientists: The challenges of online documentation. In E. Barrett, (Ed.), *Text, conText, and hyperText: Writing with and for the computer* (pp. 311–323). Cambridge, MA: MIT Press.

White, J. (1988). *Graphic design for the electronic age.* New York: Watson-Guptill.

SUGGESTED READING

Bernhardt, S. (1986). Seeing the text. *College Composition and Communication, 32,* 66–78.

Bradford, A. (1984). Conceptual differences between the display screen and the printed page. *Technical Communication, 31*(3), 13–16.

Brockmann, R. (1986). *Writing better computer user documentation: From paper to online.* New York: John Wiley & Sons.

Duffy, T. & Waller, R. (Eds.) (1985). *Designing usable texts.* Orlando: Academic Press.

Felker, D. (1980). *Document design: A review of the relevant research.* Washington, D.C: American Institutes for Research.

Felker, D., Pickering, F., Charrow, V., Holland, V., & Redish, J. (1981). *Guidelines for document designers.* Washington, D.C.: American Institutes for Research.

Galitz, W. (1986). *The handbook of screen format design.* (2nd ed.) Wellesley, MA: QED Information Systems.

Hartley, J. (1985). *Designing instructional text.* (2nd ed.), London: Kogan Page.

Heines, J. (1984). *Screen design strategies for computer-assisted instruction.* Bedford, MA: Digital Press.

Houghton, R. (1984). Online help systems: A conspectus. *Communications of the ACM,* 27(2), 126–133.

Jonassen, D. (Ed.) (1982, 1985). *The technology of text: Principles for structuring, designing and displaying text.* Englewood Cliffs, NJ: Educational Technology.

Keyes, E. (1987). Information design: Maximizing the power and potential of electronic publishing equipment. *IEEE Transactions on Professional Communication, 30*(1), 32–37.

Lichty, T. (1989). *Design principles for desktop publishers.* Glenview, IL: Scott, Foresman & Company.

Mirel, B., Feinberg, S., & Allmendinger, L. (1991). Designing manuals for active learning styles. *Technical Communication, 38*(1), 75–87.

Monk, A. (1984). *Fundamentals of human-computer interaction.* New York: Harcourt, Brace, Jovanovich.

Redish, J., Battison, R., & Gold. E. (1985). Making information accessible to Readers. In L. Odell & D. Goswami, (Eds.), *Writing in nonacademic settings* (pp. 129–153). New York: Guilford.

Rubens, P. (1986). A reader's view of text and graphics: Implications for transactional text. *Journal of Technical Writing and Communication, 16,* 73–86.

Rude, C. (1987). Format and typography in complex instructions. In F. J. Sullivan (Ed.), *Basic Technical Writing* (pp. 112–115). Washington, DC: STC, Anthology Series No. 7.

Rude, C. (1988). Format in instruction manuals: Applications of existing research. *Iowa State Journal of Business and Technical Communication,* 2(1), 63–77.

Tufte, E. (1986). *The visual display of quantitative data.* Cheshire, CT: Graphic Press.

White, J. (1982). *Editing by design.* New York: R.R. Bowker.

CHAPTER 8

PRESENTING INFORMATION THROUGH MULTIMEDIA

Francis D. Atkinson

Georgia State University

INTRODUCTION

A national furniture warehouse was slowly going out of business and could not understand why.

How could this happen? The company had the best printed catalog in the business. The catalog described each piece of furniture: its price, weight, dimensions, and color, and often included a sample picture of the item. The catalog was revised quarterly and was considered a model for smaller furniture warehouse firms. This company also had the best distribution system in the business. It could deliver items anywhere in the country within 48 hours. Finally, the company also had the best prices in the business, selling most of the furniture for less than the competition.

What was the problem and how could it be solved? More specifically, what role did technical communicators play in the solution?

Technical communicators first determined which information customers needed to make a purchase and then provided customers with that information. When the technical communicators analyzed their customers' needs, the communicators learned that customers need the following information to complete a purchasing decision:

- To see a picture of the furniture they are buying in the desired color and fabric. Although the printed catalog had pictures of the furniture, it could not show every color and fabric combination.
- To see how the furniture would look in their own rooms. Although the catalog could show how furniture might look in sample rooms, it could not show how the furniture might look in customers' own rooms.

In response to these two needs, technical communicators developed an online catalog. The online catalog is presented in the store on a personal computer using a touch screen. The online catalog uses various hypermedia programs that let customers do any or all of the following:

- See each piece of furniture in all of the color and fabric combinations available
- Scan in a photo of the room where the furniture will be placed and add the furniture to it
- Read details about the furniture, such as its weight, price, and the materials it is made of
- Order the furniture
    - □ Place an order
    - □ Apply for credit (if necessary)
    - □ Schedule delivery

This is no fairy tale vision of the future. This is the type of information system that technical communicators are increasingly involved with—systems that combine reference information with related visuals and sound and systems that integrate product information with ordering, credit, and inventory systems. As a result, customers and employees deal with only one system to meet all of their needs. Multimedia, hypermedia, and hypertext are the three technologies that make these integrated, one-stop systems possible.

This chapter explores these three technologies. Specifically, this chapter

- Explains why you should use multimedia, hypermedia, and hypertext to address your readers' needs
- Provides how-to guidelines and tips for effectively designing, using, and evaluating the effectiveness of multimedia, hypermedia, and hypertext documents

WHY YOU SHOULD USE MULTIMEDIA, HYPERMEDIA, AND HYPERTEXT

Before you begin preparing information to be presented in multimedia, hypermedia, or hypertext, you first need to understand what these media are and when you should use them. This section begins by defining these terms, then explains when you should consider using each of them.

What Multimedia, Hypermedia, and Hypertext Are

Media are the means (in this chapter, audiovisual or electronic) for transmitting or delivering information to readers. In this sense, media include pictures, films, audio recordings, video images, and computer programs.

Multimedia. The simultaneous use of more than one form of medium is called *multimedia*. Not so long ago, the term multimedia implied the use of several slide or motion picture projectors that displayed simultaneous images on several screens to present a visual "mosaic" for a large audience.

The current use of multimedia implies a much more powerful and dynamic use of sound- and image-based applications delivered by a wide variety of technologies (Bergman & Moore, 1990) for individuals as well as groups of any size. Today, multimedia applications are used to deliver every form of technical information, from stockholder reports to technical training seminars to point-of-sale marketing strategies.

A typical multimedia application uses several forms of media such as sound, graphic images, still pictures, and video sequences. These are all under the control of a computer program.

Many multimedia applications are linear in nature, allowing for little, if any, variation in the flow of the information. In many of these instances, variation is not needed. Consider, for example, the chaos that might result if a complex set of directions for completing new personnel forms was not presented in a step-by-step (linear) manner.

In other instances, readers do need to control the flow of information. Linking multimedia with other technologies makes this control by readers possible.

Hypermedia. A computer-based approach to linking information stored on different media is called *hypermedia*. "From the user's perspective, hypermedia programs are knowledge bases in which every display is a menu to others" (Locatis, Letourneau, & Banvard, 1989, p. 65). For example, hypermedia might link information in a computerized database with information on a videodisk player.

In hypermedia, information is located in *nodes*. A database, a compact disk, and a file with a computer graphic are all examples of nodes. The size of a node and the type of media it contains (such as text, graphics, music, and video) depend on the computer software used to link the media. This software is called an *authoring system*.

Links connect nodes. The success of hypermedia depends on the quality of the links. These links let users move from one node to another (Horn, 1989). Technical communicators determine which nodes are linked. The better that communicators can anticipate the links that readers will need, the more valuable the hypermedia document will be.

Hypertext. A special type of hypermedia that lets readers access text information stored in nodes and linked together is called *hypertext*. A well-designed hypertext document lets readers instantly access any part of a document and indicates the words or phrases that are linked to other text in the document.

Hypertext is becoming the primary method of presenting information online because it lets readers take advantage of the power of the computer: it lets readers quickly and easily move within documents. Moving around within other types of online documents is more difficult.

About All Three Media. One of the characteristics that distinguishes multimedia, hypermedia, and hypertext from other types of media is their reliance on computer technology. All three are developed and delivered through a computer.

Most of these *technology-based media* (a generic term for multimedia, hypermedia, and hypertext that will be used through the rest of this chapter) are designed and delivered through personal computers. A personal computer displays text, graphics, animation, still photographs, speech, music, and video in full color and, when needed, in stereo sound.

Note, too, that some hypertext programs allow for the design and delivery of documents on mini- and mainframe computers.

When You Should Use Multimedia, Hypermedia, and Hypertext

Multimedia, hypermedia, and hypertext all have unique characteristics that can be used by technical communicators. To do so, you must first understand what the communication process is and how these communication tools might improve that process. Then you can determine whether you need to use multimedia, hypermedia, or hypertext to achieve your objectives.

What the Communications Process Is. In 1954, Wilbur Schramm presented a model of the communication process that had been adapted from an earlier model by Claude E. Shannon and Warren Weaver (Schramm, 1954, p.116). Schramm's model, presented in Figure 8.1, emphasizes the two-way relationship between senders (communicators) and receivers (readers).

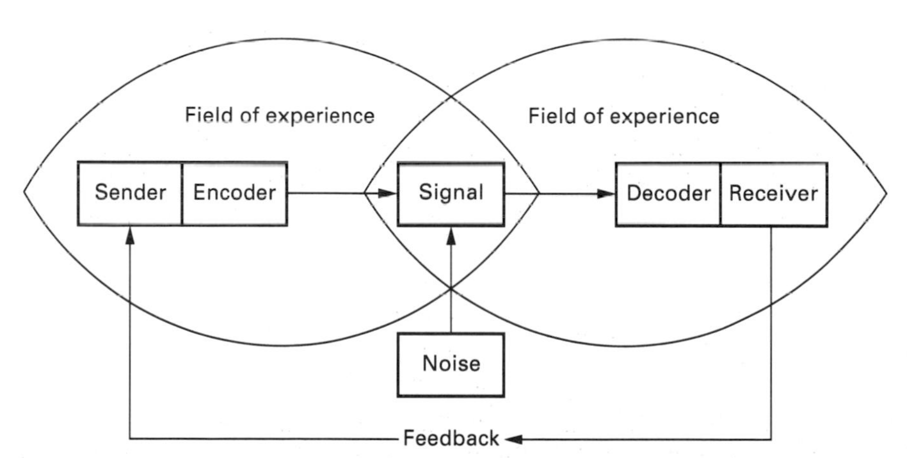

FIGURE 8.1 Schramm's Communication Model (Schramm, 1954, p. 116)

As the *sender*, you encode—or translate—the information to be communicated into symbols (such as words and pictures) and then select the most appropriate signal to send these symbols to readers. The signal can be a medium or media through which you transmit information, such as print, sound, still pictures, video, and computer graphics.

As the *receivers*, readers are the group for whom you prepare information. Readers must decode—or translate—the symbols you used to communicate the message. As noted in Chapter 1, the better that communicators prepare information for readers, the better readers will understand it. In other words, you should prepare information in terms that readers can relate to.

Three factors—common experience, noise, and feedback—play key roles in the success of the communication process.

The role of common experience in the communications process. Both senders and receivers possess a "field of experience." The greater the overlap of this field of experience, the more successful the communication is likely to be. That is, the more senders and receivers have in common, know (or can find out) about each other, the more likely it is that successful communication will occur. Your "field of experience" as a technical communicator is critical in this process, particularly in the selection of the most appropriate "signal" (communication medium).

For example, consider the technical communicator who learns that her audience has limited reading skills. Video or audiotape would probably be the best communication medium for reaching this audience. But this communicator has a limited field of experience with video and audio. Because she lacks skills in these areas, she chooses to communicate with her audience through print. She makes a strong impression on her management because printed materials usually cost less than video and audiotapes, but she fails to communicate with her readers because they cannot read the printed materials.

In this instance, the communicator should have been aware of research that indicates people often have a preferred medium for sending and receiving information. She should have selected the medium that her audience preferred, not the one she felt most comfortable with. She needed to expand her field of experience. Similarly, you, too, might need to expand your field of experience to meet your readers' needs.

The role of noise in the communications process. *Noise* is anything that distorts or gets in the way of the intended message (see Figure 8.1). For example, inadequate lighting in a room or poor color contrast on a computer display can distort the intended message. In the same manner, highly charged emotions can distort the intended message.

As a communicator, you need to identify and reduce (or eliminate) potential sources of noise that might distort your intended message. For example, a technical communicator was asked to prepare a series of color slides and an audiotape showing employees how various departments would be affected by a company "downsizing" (an effort to reduce its size by eliminating jobs and departments). While working on this task, the communicator discovered two possible sources of noise:

- The information would be presented in the company cafeteria, a large-windowed room that could not be made dark. Slides would not show well in this room.
- The topic had already generated considerable anxiety among employees. Many were afraid that they would be "downsized" out of jobs.

In response, the technical communicator made the following recommendations to make sure that the information was successfully communicated:

- Use full-color overhead transparencies instead of slides because transparencies are less dependent on a dark room than slides.
- Have an officer of the company deliver the presentation and answer questions from the audience during the presentation.

Both recommendations were accepted—well, more or less. Transparencies were used and an officer of the company was available to answer questions after the presentation.

The presentation itself was audiotaped, however. The company chose this medium to ensure against another type of noise: inconsistency. The presentation was going to be given to several groups over a two-week period and the company's executives wanted to make sure that everyone heard the exact same message. Only by taping the message could the officers have this assurance.

Similarly, in the process of analyzing a communication problem, you might identify physical and emotional sources of noise that can be addressed through a combination of media—multimedia.

The role of feedback in the communications process. *Feedback* is the part of the communications process in which the receiver sends a message back to the sender, telling the sender whether the message was received and whether the receiver understood it (See the feedback loop in Figure 8.1).

Feedback lets you adjust the communication process to meet readers' needs. All communication can provide for some form of feedback. The more that senders can interact with receivers, the more the opportunities for feedback.

For example, a sales manager can see the results of a presentation given in a small seminar to a few sales representatives and adjust the presentation accordingly. From observing sure nods to confused faces, the manager can see whether the presentation is successful. By listening to the questions and comments received, the manager can hear whether the presentation is successful.

You can also get feedback about printed materials, but you rarely receive it during the reading process. You can attach a reader response card in the brochure to receive feedback. You can schedule a test with intended readers to determine whether the materials are working as intended or need to be adjusted (see Chapter 11 for more information on testing).

Similarly, one of the advantages of hypermedia and hypertext is that they let you anticipate communication problems and develop several strategies for addressing them. Through interaction with hypermedia and hypertext documents, readers can select the communication strategy that most effectively addresses their needs.

The role of active involvement in the communications process. The more senses involved in communication, the more effective it will be. For example, most readers learn more quickly and retain what they learn longer if they both see and hear the message.

Communication can be even more effective if receivers have an opportunity to see, hear, and do something related to the message. This implies that the best communication requires that readers (receivers) be actively involved in the communication process and "do" something with the information being presented.

Consider, once again, the solution for the example presented at the beginning of this chapter. The audience can see and hear about the product, as well as "use" the information. The hypermedia catalog lets customers consider design choices such as the color of furniture and its placement in a room.

Actively involving readers also reduces noise because readers can experience for themselves exactly what the communicator means.

When to Use Multimedia, Hypermedia, and Hypertext. You should choose a technology-based medium, such as multimedia, hypermedia, or hypertext, when it most efficiently and effectively communicates your message within the budget and time available to develop the materials. That's a loaded statement, so consider it part by part.

Choose a nonprint medium when it most effectively and efficiently communicates your message. That means, choose a nonprint medium to address such concerns as readers' preference for nonprint materials, to minimize noise and to provide feedback in the communication process, and to provide an opportunity for readers to "hear and do" as well as see in words.

From a growing body of research we can draw the following generalizations about the effectiveness of communication through technology-based media. Using technology-based media can achieve these objectives:

- Increase motivation
- Reduce learning and communication time
- Reduce costs
- Provide consistent quality
- Provide a safe environment for learning and communicating
- Provide greater involvement by the audience

Technology-based media are especially useful for employee communications, training, and sales materials. For example, as mentioned earlier in the example on downsizing, consistency among presentations of the same material is sometimes important. Recording the message through audiotape, videotape, multimedia, or hypermedia provides such consistency. Without consistency, different audiences might hear different versions of the same message and, when they compare what they heard, noise results (Locatis & Atkinson, 1984). Similarly, technology-based media can provide a unique level of feedback not possible in print. For example, a computer-based typing course begins by asking users to identify their education level and their typing experience. Based on these two factors, the computer adjusts the course so users receive information suited to their education and experience.

Technology-based media also provide an opportunity to show and tell; print media can only tell. Marte Pendly, Director of Corporate Communications for Mervyn's Department Stores, commented:

> The media we use most are meetings, print, video, and audio in that order. And very often, we go with all of them together. (Weinberg, 1991, p. 11)

Pendly commented that Mervyn's produced a video to tell employees how to develop effective in-store displays.

Finally, technology-based media provide an opportunity to "do" as part of the learning process. For example, many software applications contain tutorials. These tutorials not only tell users which tasks they can perform with the software, but also let users perform these tasks. Consider one tutorial for a word processing program that lets users enter, move, change, and format text. Consider another tutorial for a computer operating system that lets users start and stop jobs. In these instances, users are not actually performing these tasks; they are performing simulations of the tasks.

When Development Fits Within the Budget and Time Available. As mentioned in Chapter 2, technology-based documents, such as multimedia, hypermedia, and hypertext documents, usually take more time to develop than print-based documents and cost more, too.

Technology-based documents take more time to develop than print-based ones because development involves more than the writing, and production processes are more complicated than for print. For example, multimedia and hypermedia documents usually consist of several smaller documents, some written for video, some for audio, some as computer-based text, and some as computer programs. Although these documents are written separately, the writing must be coordinated to make sure that information is consistent and avoids unnecessary repetition. Production of technology-based documents, too, is more complicated than production of print documents. For example, video usually requires a week of production and another week of editing (preparing the tape for viewing) for each thirty minutes of finished videotape. Production itself is complicated because it requires a large team of people including a director, sound and lighting specialists, actors, and stage crew.

Production is further complicated by the process of linking all the documents. All of the links must be programmed and tested to make sure that they work. Finally, materials must be stored on computer disks and duplicated, a process that adds additional time and complexity.

This complex writing and production process is not only lengthier than that needed for print but is also costlier.

Whether you can afford this additional time and funding depends on the guidelines established for your project. Chapter 3 explained how to establish a budget and a schedule for a project. It mentioned that, in some instances, you can determine the budget and schedule based on your plans. In other instances, the budget and schedule are already set for you and you must make choices that ensure you meet the budget and schedule. These choices must also be justified by the additional benefits that readers receive because you chose technology-based media.

Although rules of thumb for estimating the cost and time needed to produce technology-based documents exist, these rules of thumb are usually inaccurate. Technology-based media offer so many production choices that each choice significantly affects the cost and time needed to develop these documents.

The most honest answer to questions about the relative costs and benefits of technology-based communication, then, is that "it all depends." It depends upon such factors as the skill and experience of the technical communicator, the skill and experience of the intended audience, the complexity and length of the message to be delivered, the type of multimedia selected, the type of computer software and hardware to be used, and whether information from existing programs can be incorporated into the new one.

Rather than following a rule of thumb to determine whether the cost is affordable, you need to consider some guidelines and make a judgment about whether a technology-based medium addresses the needs of your audience. These guidelines should prove useful as you mix your professional judgment and experience to make the best decisions about design, development, and delivery of documents.

Your skills and experience. Your previous experience with technology-based documents—both as a user and as a communicator—is one of the most important ingredients in determining the costs and benefits of a particular solution. The more skill and experience you have with these technologies, the more effectively and efficiently you can develop documents for them. Consider your experience as a new user of a word processor and as an experienced user. With experience, you can produce documents in less time than without it. At the same time, you probably use special features of the word processor that you did not use your first few times. Similarly, you can produce documents more efficiently when you have experience with an authoring system. You can also make more effective use of the special features of the technology. For example, the first online reference manuals used menus and "search strings" (words that users typed in and the system located) to help users move within the document. More recent hypertext references let users move from place to place merely by clicking on a word with a mouse.

The audience's skill and experience. The skill and experience of the audience affect two areas:

- To what extent does the intended audience have skill and experience with the content and objectives of the information being communicated? If the information is a review of something the audience thinks it already knows, you might use technology-based media to provide feedback. For example, you might begin an annual presentation of chemical safety rules with a pretest. From the score, readers can determine how much they really know to determine how much attention they must pay. You might also tailor the presentation so readers only have to go through the sections presenting material that they had difficulty with on the test.
- To what extent does the intended audience have skill and and experience with the medium used to deliver the message? For example, will the audience know how to use a mouse, or must they be told? If the medium is new to the audience, at the very least, they need to learn how to use it. At the most, the audience might resist it altogether.

  Consider, too, whether the audience knows how to use special features of the medium. For example, will the audience know how to read complex visuals, such as engineering diagrams? You might need to provide training on these aspects of the medium in addition to presenting the information.

Complexity and length of the message to be delivered. Generally, the more detailed, complex, and technical the information, the more time needed to design, develop, and produce the document and the more time needed by the audience to go through it. For example, it required more development time to design the hypermedia catalog described at the beginning of this chapter than it did to develop the printed one. But the additional investment was worthwhile because it better met the needs of the audience.

Technology selected. In general, complex media formats, such as well-designed hypermedia involving video and audio, require more development time than less complex media, such as linear video and audio recordings. It is also a general rule of thumb that documents with several types of media take more time to develop than documents with one medium. Consider the extra development time it takes to combine just the right graphics, still pictures, and video sequences to communicate a company's new initiatives and rules regarding a recycling campaign. If this were a brochure, the communicator would need only to select a photo or a drawing and would have less material to review, much less produce.

Type of computer software and hardware. Care must be taken when selecting the "right" computer software and hardware for developing and delivering technology-based documents. Matching the current and future needs of your organization with the features of a computer system, such as ease of use, flexibility, and power, is a major undertaking. With new and exciting software and hardware announced on what seems to be a daily basis, you can become easily bedazzled by the "bells and whistles" of the technology. But before you are seduced by a bell and buy a whistle, make sure you really need these features. For example, when selecting a display for a computer, determine how sharp the picture really needs to be. You might save hundreds of dollars if you find that you can make do with lower resolution. Similarly, you might find that an authoring system can let you convert information from a personal computer to a host—or mainframe—computer. If you have no need for the conversion and do not anticipate such a need, you do not need this feature.

Availability of existing materials. Can you incorporate some existing video and sound sequences or computer graphics into your document? If you can, you might be able to reduce the cost of production.

Technical communicators sometimes feel compelled to develop all new material for each new document they write. When the information is not new, however, and can be used as written in the technology-based document, you save both time and money developing the new document. Because the time needed to develop technology-based documents is long and the cost high, reusing information is often the only way to make these documents affordable.

For example, you might discover that a coworker developed several computer graphics showing common safety hazards in the workplace. With minor changes, you can use these graphics in your hypermedia-based orientation program for manufacturing employees.

When using existing information, consider the following:

- The information must address the objectives set for the new document. (Chapter 3 explains how to establish objectives for a document.)
- The organization you are developing the materials for must already hold the copyright on the reused material or must be able to obtain it. For example, if you are a contract writer and developed information for Company A, you cannot use it in material you are developing for Company B because Company B does not hold the copyright on the information and probably cannot obtain it. (See Chapter 12 for a detailed discussion of copyright laws.)

Reusing information is an instance of generic writing, which was described in Chapter 5. You can promote the reuse of information by maintaining a development database of information modules. For example, you might establish a corporate-wide computer graphics library from which everyone could get relevant graphics for technology-based documents. The company logo, copyright notices, credits, borders, and many other generic graphic displays need only be created once and then changed as necessary.

HOW YOU CAN DEVELOP MULTIMEDIA, HYPERMEDIA, AND HYPERTEXT DOCUMENTS

Developing information for technology-based documents involves the following:

1. Determining which medium to use for each piece of information.
2. Preparing the information for each medium.
3. Producing the information.
4. Linking the information.

Determining Which Medium to Use for Each Piece of Information

If you are preparing a multimedia or hypermedia document, you can choose from a variety of media. As mentioned earlier, multimedia and hypermedia are actually computer tools used to link information produced in more basic media. Following are some of the media more commonly used in technology-based documents and suggestions for when to use them:

Medium	When You Should Use It	Example
Text	For information that is verbal only and has no visual component.	Explanation of how to use a command.
Sound	For information that needs to be heard, either because hearing the information gives it more credibility or because the audience must know the sound itself.	The sound of an emergency alarm that the audience is expected to respond to.
Still visual	For information that cannot be adequately described with words. (See Chapter 6 for an explanation of when to use visuals.) You can use either computer-generated visuals or photos. Choose the type of visual that best meets the needs of your audience and, at the same time, can be generated within your budget and technical limitations.	the location of all emergency exits in the building.
Moving visual	For visual information that involves movement. You can use either computer-generated animation or moving video sequences.	An explanation of a manufacturing process.
Combinations of media	For most types of information. Usually, no single medium fully achieves your goal. In those instances, combine them as needed.	A narrating sequence that describes a manufacturing process and that has key words displayed on the screen.

After you determine which medium to use for presenting each type of information, you need to record this information on a storyboard. (Chapter 2 explains how to develop a storyboard.) The storyboard shows how the information appears to readers; a single board might show video, audio, and text components if all three are being used.

By preparing a storyboard, you ensure that the information produced in the different media is coordinated. This is especially important if different people write and produce the video, audio, and text parts of the document.

Preparing the Information

Preparing information for nonprint media is similar to preparing printed information. After you develop the storyboard, you write the information and have experts review it.

As you write the text and design the visuals, however, you need to consider the unique aspects of video, audio, and computer text that distinguish them from print. Because of these differences, the experience of reading (or viewing) changes. This, in turn, affects how people get information from these media and how much information they retain. In other words, the literacy skills for technology-based media differ from those for print media (Marsh, 1983).

Key characteristics that make these media unique are:

Medium	Characteristics
Text	Research shows that people read information online only 75% as fast as they read it on paper (Rubens & Krull, 1985). Therefore, do not crowd information on a screen. Readers do not read online text in a linear manner (from beginning to end). Instead, they read the parts that interest them, possibly ignoring many other parts of the document. Reading text online is like a conversation, of sorts, between a reader and a computer. The computer "asks" readers what they would like to read, and readers type a response. The computer then presents the selected information. Because of this dialogue-like experience, you use a more conversational approach when writing for online documents (Carliner, 1990; Jonassen, 1985).
Sound	Listeners listen four times faster than speakers speak, which leaves listeners with many opportunities to tune out a presentation. You must make a special effort to maintain interest. Most likely, this effort involves the use of visuals. Spoken language is different from written language. Spoken language is more colloquial; it uses contractions, incomplete sentences, and common expressions—tools that you are usually told to avoid when writing. You need to learn new rules of grammar when writing scripts (Carliner, 1987).

(*continued*)

Visuals (still or moving)	As the old saying goes, a picture is worth a thousand words. Therefore, let your visuals do the speaking and let words "support" them. When using a visual, let it—not words—be the primary means of communicating with readers. Use narration (sound) and text only to explain those parts of the visuals that readers might not understand.
Computer	The computer lets readers move about a document at will. As long as the document contains the links that readers want, readers can move from part to part as they choose. In contrast, readers of printed materials must either skim pages or go through an intermediate device like a table of contents or an index to move within a document.

The following guidelines suggest how you might use these characteristics of technology-based media to help your readers:

- Carefully control the flow of events. Although one of the advantages of technology-based media is that they provide readers with more freedom to move about, you might not want to provide this freedom from the start. For example, before letting people use an online information service, you might want them to view the rules for using it.
- Make sure that readers see certain information. You can design the document so that all readers must go through the introduction before choosing another part of the document to see. You can either prevent readers from moving forward until they read the section or design the links in such a way that all readers are guided to the information of interest.
- Make sure that readers know the structure of the document. Readers receive the most value from a document when they know what it contains and how to get to different parts. You can provide this information in the following ways:
  - Place a menu of information available at the beginning of the document.
  - Include an "About this Document" section.
- Maintain reader interest by introducing novel or unexpected events. The novelty should be relevant so that it does not distract readers from the task at hand. Animation and sound cues are possible novelties.
- Remind readers of relevant prerequisite information and, when possible, provide the links. For example, if this book were a hypertext document, this chapter would be "linked" to chapters 2, 3, 5, and 6 because it refers to prerequisite information in those chapters.
- Provide clear prompts and cues so readers know when and how to respond to requests for information from the system. For example, if you want readers to select an option on a menu, say so, using those words.

- Write in conversational language. Use contractions, phrases, one-word sentences, and colloquial terms that characterize speech. When possible, use language specific to the audience. For example, if are writing a hypermedia document to be used by doctors, you would use the specialized vocabulary of medicine (Heckel, 1991).

Producing the Information

After the information is written and reviewed, it is produced. During production, the ideas you have developed are converted into computer graphics, audiotape and videotape sequences, and online text.

As mentioned earlier, the production process is complex and beyond the scope of this chapter, but you should be aware of what happens during this process so you can write materials that can be produced with a minimum of difficulty. The following are the parts of a project that need to be produced and the considerations for producing them.

Medium	Production Considerations
Text	Text is the easiest of the media to produce. Production involves programming in the links with other documents and formatting the text for presentation online. When programming in the links, present information in the smallest units possible. Ideally, all of the information on a specific topic should be found on a single screen. If the topic requires more than one panel, consider separating it into more than one topic. Problems with text are easy to resolve because you can quickly and easily change text. See Chapter 7 for a list of considerations for formatting online information.
Audio	Audio production involves recording the narration, adding sound effects and background music (if desired), and "engineering" the sound to eliminate unnecessary flucuations in sound quality. Specific considerations include ■ Making sure each word is clearly spoken. ■ Avoiding distracting background sounds. The considerations for audio recording are the same as those for the audio part of a video sequence. Note, however, that audio is difficult to change because a change often involves re-recording and re-engineering. During re-recording, the narrator often sounds different.
Visuals	The production of visuals involves turning ideas into pictures, whether they are video sequences or computer graphics. Chapter 6 explains the considerations for producing graphics.

Consider, though, the complexities of video production. Some of the issues you need to be aware of include the following:

- Location. Informational videos are usually shot in a place that realistically looks like the environment described in the script. Sometimes, the environment can be recreated in a studio. This involves the expense of building a set, but has the advantage of providing greater control over the technical aspects of production. For example, in a studio the producers do not need to worry about unnecessary background noise.

  Sometimes the video must be shot "on location," which is expensive and time consuming because it requires one or more visits before shooting to set up. It also causes disruption to those people who regularly use that location as a workplace.
- Talent. Consider who narrates or performs in the video. Do you use real people, who provide credibility but might not have acting skills, or actors, who are not known to the audience but can skillfully portray their experience? Also consider the relative expenses of using professional actors or employees.
- Production values. *Production values* refer to the expense and attention paid to such details as lighting, sound, and costuming: details that determine the overall quality of the production. The higher the production values, the more realistic—and costly—the video. Most audiences are used to the high (expensive) production values of broadcast television, but most industrial video budgets are limited. Production values can be scaled back to accommodate budget needs, but this should be considered when planning the project.
- Post-production. *Post-production* is the process in which the various video sequences are edited into a single production. Sequences are usually shot out of order to reduce cost. All of the scenes that occur at one location are usually shot at the same time, even if they do not occur consecutively in the script.

  During post-production, other aspects of the presentation are also addressed. Background music, sound effects, titles (the title frame at the beginning), and credits (the list of who did what, usually shown at the end) are added. The video is also engineered to eliminate unnecessary variations in sound and visual quality.
- Production time. Because production involves many people, it must be scheduled a month or more in advance to make sure everyone—from the actors to the lighting crew—are present.

  Video is difficult to change because a change often involves re-recording and re-engineering, so make sure you record your video correctly the first time.

Linking the Information

After the various components of the document are produced, you link them by using an authoring system. At a minimum, this system should provide you with control over all the media used in the presentation.

The authoring system might also let you provide standard information on each screen, such as how to get help and how to end the presentation. This feature should save production time. Finally, the authoring system might let you divide the screen into several sections, so you might be able to show a computer graphic in one section, related text in another, and a related video sequence in a third. These sections are called *windows*.

Your primary concerns when linking information are technical ones—making sure that the document works as intended. Concerns in this area include:

- Making sure that the information that is supposed to appear actually does appear. That is, you need to test every link you programmed to make sure it works properly.
- Making sure that the various media work without problems. That is, if a video sequence is included, make sure that it runs from beginning to end without difficulty.
- Making sure that the system properly accepts and acts on input. For example, if readers are supposed to be able to search for terms, make sure this function actually works. Similarly, if the document includes a tutorial with exercises, make sure that the exercises provide the right feedback. For example, if readers type an incorrect response, the system should tell them so.

Testing the document with readers to make sure it works as intended is one of your most important responsibilities. (See chapter 11 for a discussion on testing documents.)

SUMMARY

This chapter explained how to produce documents for technology-based media. First, it introduced you to three types of technology-based media:

- Multimedia
- Hypermedia
- Hypertext

Next, this chapter suggested uses for these media. When choosing a medium, determine how it best addresses the communication problem at hand. You can do this by considering several factors:

- The communication model and how the proposed medium matches the needs of the audience, reduces noise, provides feedback, and involves readers.

- The budget and schedule for your project. Technology- based media usually require more time and funds to produce than printed documents, but the additional investment can be justified economically because the information is more useful to readers. Other budget and schedule considerations include these factors:
  - Skill and experience of the technical communicator in producing technology-based documents
  - Skill and experience of the intended audience in using technology-based documents
  - Complexity and length of the message to be delivered
  - The type of computer software and hardware to be used
  - The availability of information from existing programs that can be incorporated into the new one

Last, this chapter presented a four-step process for developing documents for technology-based media:

1. Determine which medium to use for each piece of information.
2. Prepare the information for each medium by capitalizing on its strengths.
3. Produce the information.
4. Link the information with an authoring system.

LIST OF TERMS

authoring system Software used to develop hypermedia documents and link nodes together.

feedback Part of the communications process in which the receiver sends a message back to the sender, telling the sender whether the message was received and whether the receiver understood the message.

hypermedia Computer-based approach to making links among information stored on media that are controlled by a computer.

hypertext Special type of hypermedia that lets readers access text information stored in nodes and linked.

links Connection of nodes in a hypermedia document.

media Means (in this chapter, audiovisual or online) for transmitting or delivering information to readers. In this sense, media includes pictures, films, audio recordings, video images, and computer programs.

multimedia Simultaneous use of more than one form of media. These days, it refers to sound- and image-based applications delivered with a wide variety of technologies.

nodes Information linked by a hypermedia program. This information can come from many sources, such as a database, a compact disk, and a text file.

noise Anything that distorts or gets in the way of the intended message.

post-production Process in which various video sequences are edited into a single production.

production values Expense and attention paid to such details as lighting, sound, and costuming.

receiver Person in the communications process who receives a message.

sender Person in the communications process who sends a message.

technology-based media Generic term used in this chapter to refer to multimedia, hypermedia, and hypertext.

windows Sections of a computer screen that can be used for different purposes.

REFERENCES

Bergman, R. E., & Moore, T. V. (1990). *Managing interactive video/multimedia projects.* Englewood Cliffs, NJ: Educational Technology.

Carliner, S. (1987). Audiovisual words: The scriptwriter's tools. *Technical Communication, 34*(1), 11–14.

Carliner, S. (1990). Elements of editorial style for computer-delivered information. *IEEE Transactions on Professional Communication, 33,* 38–45.

Heckel, P. (1991). *The elements of friendly software design: The new edition.* Alameda, CA: SYBEX, Inc.

Horn, R. E. (1989). *Mapping hypertext: The analysis, organization, and display of knowledge for the next generation of on-line text and graphics.* Lexington, MA: The Lexington Institute.

Jonassen, D. H. (Ed.). (1985). *The technology of text: Principles for structuring, designing, and displaying text* (Vol. 2). Englewood Cliffs, NJ: Educational Technology.

Locatis, C., Letourneau, G., & Banvard, R. (1989). Hypermedia and instruction. *Educational Technology Research and Development, 37,* 65-77.

Locatis, C. N., & Atkinson, F. D. (1984). *Media and technology for education and training.* Columbus, OH: Charles E. Merrill.

Marsh, P. O. (1983). *Messages that work: A guide to communication design.* Englewood Cliffs, NJ: Educational Technology.

Rubens, P., & Krull, R. (1985). Applying research in document design to online information. *Technical Communication, 32*(4), 29–34.

Schramm, W. (1954). *Mass media and education.* In N. B. Henry, (Ed.), *Fifty–third yearbook of the National Society for the Study of Education* (Part II). Chicago: University of Chicago Press.

Weinberg, G. (1991, October). PR and tech writing have common techniques. *Active Voice,* p. 4.

PART III

TECHNIQUES FOR EDITING INFORMATION

CHAPTER 9

REMEMBERING THE DETAILS—MATTERS OF GRAMMAR AND STYLE

Fern Rook

Consultant, and Fellow,
Society for Technical Communication

communication	enhances	understanding
(subject)	(verb)	(object)

For technical communicators, many writing tasks are matters of dispensing information or giving directions. As a result, your approach to sentence structure will necessarily be more pragmatic than that of novelists, poets, and philosophers. Whether you are a seasoned technical communicator or a student of technical communication, you must approach a writing assignment as a building process. Each of you must first determine what you need to accomplish. Then you must gather the necessary material, decide how to present it, and determine the reading level of your readers. Last you must decide on the kinds of sentences you must use to comply with the purposes of a specification or document. To select the appropriate kinds of sentences, you must determine whether you are writing to give information or to direct your readers to perform certain tasks in particular ways. Only after you know the purpose of a document, the readers, and the right kinds of sentences, can you proceed with the writing task.

Previous chapters in this book have explored ways to understand your readers, to plan your projects, and to prepare to write. This chapter explores English syntax—how sentences are constructed—and includes the entire field of English grammar and usage. I will review those aspects that I think may be most helpful to you as technical communicators:

- Simple sentence syntax
- The English verb system
- Modifying elements in English sentences

SIMPLE SENTENCE SYNTAX

English sentence structure is based on syntax, the relation of words to other words in a sentence. You already know how to make statements, ask questions, and make commands. You know that when you want to make a statement, the subject comes first, the verb second, and, if the verb takes an object, the object comes third. This knowledge is basic to forming English sentences. "Children feed puppies" says something quite different from "Puppies feed children."

Technical communicators must produce material that is easily read and understood. In writing specifications you often must detail not only the format of a document, but also prescribe sentence type and vocabulary limits. You are most often asked to use simple sentences, and these must often be limited to fewer than 15 words. *Reader's Digest* is a good example of simple sentences directed to general readers. Actually, simple sentences constitute about 90% of all writing done, both literary and informative.

Grammatically speaking, simple sentences contain only one *clause*. Some contain only a subject and a verb, but simple sentences often contain an object, an indirect object, or a predicate nominative, as the following examples illustrate.

> Bosses complain. (*Subject, verb*)
> Bosses make policy. (*Subject, verb, object*)
> Bosses are fair. (*Subject, verb, predicate adjective*)

In discussing simple sentence syntax, let's first examine subjects and objects of sentences.

Structures that Comprise Subjects and Objects of Sentences

Nouns. In simple sentences, subjects and objects are usually nouns and pronouns. As you may recall, nouns are usually defined as being the name of a person, place, or thing. This definition is far too simplistic, however, because it omits the fact that nouns also can name an idea or quality. The simplistic definition can cause confusion, as was the case with a student I once had. "What is 'death'?" he asked. "It isn't a person, place, or thing." So it was necessary to give him a pragmatic answer. "You can tell it is a noun because you can use the articles before it, *a* death, *the* death, and you can make a plural, *deaths.* It can be the subject of a sentence or the object of a verb. It can be used everywhere a concrete noun can be used." "Oh," he said. To avoid this confusion, make sure you include "idea and quality" in your definition of nouns.

Personal Pronouns. Today's technical communicators use more personal pronouns than were allowed twenty years ago. The personal pronouns we use as subjects—pronouns like *I, we, he, she, it,* and *they*—not only lend a sense of reader participation, but also allow us to use more active verbs than were formerly acceptable in technical or business writing. We must use the objective forms *me, him, her,* and *them* when they are objects of verbs, verbals, or prepositions.

We should never use pronouns simply to avoid passive verbs. The word *we,* for example, might indicate an opposing *they*. There is nothing wrong with saying, "The job was estimated to take 54 hours, but was completed in 48." Introducing personal pronouns to avoid passive verbs can inject subjects that might influence the sense of the sentence. "We estimated that the job would take 54 hours, but we completed it in 48" could easily lead a reader to assume that someone has estimated poorly rather than that the group had worked effectively.

Usage Note. As you know, singular personal pronouns designate the sex of the referent, and the pronouns most often used to designate an unknown referent are *he, his,* and *him*. For example:

> If anyone calls, tell him I am busy.
> Tell the applicant to leave his resume.

Many people feel that such use of masculine pronouns tends to relegate women to an anonymous status, even though grammarians assure us that *he, his,* and *him* have always been used to indicate unknown referents. Women know that *he, his,* and *him* are, nonetheless, masculine pronouns and that the connotations of masculinity are very strong.

Technical and business writers ought, therefore, to consider ways to avoid sexist language. Today, the names of many occupations have been changed (except in the military), so careful writers are likely to refer to a policeman as a police officer or a postman as a letter carrier or postal worker. In avoiding sexist terms and pronouns, remember that job titles ought to refer to the task, not to the gender of the person doing the work.

In addition, consider the four following suggestions for avoiding sexist pronouns:

1. Use plural nouns:

   - *Instead of:*
     Tell each applicant to leave his resume.
     Use: Tell the applicants to leave their resumes.
   - *Instead of:*
     A pilot must update his medical records periodically.
     Use: Pilots must update their medical records periodically.

2. Use passive verbs. (Disregard any computer beeps that tell you that passive verbs are wrong in all cases.):

   - *Instead of:* A pilot must upgrade his (her) medical record.
     Use: A pilot's medical record must be updated.
   - *Instead of:* A doctor must spend his (or her) first year as an intern.
     Use: A doctor's first year must be spent as an intern.

3. After indefinite personal pronouns, use *they.* Although *anybody, anyone, somebody, everybody, everyone,* and *someone* have historically been classified as singular, it is now generally thought permissible to use *they* or *them* to refer to the subject:[1]

   - If anybody (anyone) calls, tell them I am busy.
   - Somebody (someone) left their keys on the counter.
   - Everybody (everyone) must confirm their reservations.

4. Consider rewriting questionable sentences. You can often avoid using pronouns altogether.

---

1. Strict grammarians would maintain that this usage is still not correct, although it is widely used in spoken English. *Eds.*

Verb Forms Ending in -ing. Verb forms ending in -ing are often used as subjects and objects of sentences. Grammar books call these *gerunds*.

For example:

Running is a good exercise.

Gerunds do not necessarily perform like other nouns, however. Many do not have plurals, although we can think of some: The *showings* are at 8 and 10 o'clock; his *comings* and *goings* were noted by his neighbor.

Gerunds formed from transitive verbs can also take objects:

Landing an aircraft is a precise maneuver.
Cashing a check requires identification.

Infinitives. Infinitives are formed by the word *to* and the plain form of a verb: *to test, to improvise, to generate.* Although infinitives are verb forms, they name only the action of a verb. They do not indicate any person or thing that performs an action, nor do they show any sense of time. They are used in sentences wherever nouns can be used.

Both transitive and intransitive infinitives can appear as subjects of sentences. The object of an intransitive infinitive then becomes part of the subject:

To locate the laboratory equipment was the first step.
To read five books was her latest English assignment.

Usage Note. Although no rule of grammar states that you cannot split infinitives, some people object very strenuously. My own observation is that there is little difficulty in placing adverbs outside intransitive infinitives (those that do not take objects): *to walk fast, to rain hard, to think objectively.* When the infinitive is transitive, however, we may sometimes wish to split it:

The president appointed her to *carefully investigate* alleged inconsistencies in the accounting procedure.

Sometimes complex sentences allow no other choice:

Writers who fear breaking the rules of grammar keep hoping to find a way *to conveniently and consistently solve* controversies (that) they often encounter when managers and editors evaluate their work.

We obviously cannot separate the infinitive from its object with the adverbial modifiers, nor can we interrupt the close relationship between "controversies" and the modifying restrictive clause, "that they often encounter." We definitely ought not place the adverbs before the infinitive because they might be understood to modify "to find a way."

Sometimes it is simpler and better to split an infinitive.

Verbs. Occasionally we use verbs as subjects and objects of sentences:

His first *try* was not successful.
The *buy* was the largest on record.

Clauses. Clauses are groups of words that contain a subject and a verb. Certain clauses can be subjects and objects of sentences:

Whoever she chooses will be acceptable.
Whatever method he recommends will be carefully tested.

Flexibility of Parts of Speech

Many words function as nouns or verbs depending on where we use them in sentences. Leaf through your dictionary and you will find hundreds of examples such as *daydream, exhaust, exercise, farm, groan, hatch,* and *host.* Words that function both as nouns and verbs have been a characteristic of English for hundreds of years. Those who arbitrarily rule that we cannot use a noun as a verb or a verb as a noun are unfamiliar with basic English *grammar* and *usage.* No one now believes that *access* functions solely as a noun. *Input* is used both as a noun and a verb, although the verb has gained a past tense that grates on some ears: *inputted.* In every technological field you will find old words being used in new ways and new words coined as the need arises.

The most common practice is simply to move a noun into a verb position. For example, when the first satellite was put into orbit, the newspapers made a new verb to describe it: "orbited."

The main problem is not to call more attention to your coinage than to your message, as the Beef Council did when it asked us to follow the latest trends in beef "menuing."

THE ENGLISH VERB SYSTEM

Having briefly examined the kinds of words and phrases we use as subjects and objects of simple sentences, let us look at the heart of English sentences—the verb system. Verbs are the most dynamic words in sentences. They indicate existence, actions, and what happens to people and things:

The sky *is* cloudy.
The car *runs* well.

Accidents often *occur* during rush hour.
Fatal accidents *are* often *caused* by drunk drivers.

We classify verbs according to how we use them. You are, no doubt, familiar with the terms *active* and *passive* voice. Grammars do not generally discuss phrasal verbs, but you need to know something about them. And you need to know about the tenses of verbs, since they are time indicators.

Active Voice

When the subject of a sentence is acting, performing, being, or becoming something, verbs are in the *active voice,* as in these examples:

The bank *closes* at 5 o'clock.
Guard dogs *patrolled* the yard.
Sarah *is* small for her age.
The park *will be* open in June.

Technical communicators are generally urged to prefer the active voice except in the following situations.

Passive Voice

We use the *passive voice* when we write about matters where an agent (the person or thing who performs the action described by the verb) is unknown or is not important. Sometimes we either do not care who performed an act or had rather not say. We can also use the passive voice to shift emphasis. We might say, "An earthquake devastated Armenia," using the active voice, but we can shift the emphasis by saying, "Armenia was devastated by an earthquake." Although the former version is more economical, the latter version provides a different emphasis, and thus shifts the sense of the sentence. Technical communicators must choose the most effective way to present their material for each situation.

Sentences can correctly contain either active or passive verbs. Passive verbs are those using past participles with some form of the verb *to be.* The subject is being acted upon as in:

He *was fired.*
Voters *were registered.*
Products *are being tested.*

Passive verbs are preferable when we speak in general terms rather than in specific instances:

Smallpox *has been eradicated* from the world.
Medical researchers *are criticized* for using live animals in experiments.

Usage Note. Some people have been led to believe that using passive verbs is wrong. You will even find computer editing programs that flag passive verbs. However, computers cannot make rules of grammar, although some programmers seem to try. You and I know that there is nothing wrong with saying, "The plan was submitted to the board for approval early in May." To say, instead, "We submitted" could introduce an unidentified subject that serves no purpose other than to avoid using a passive verb. You must not blame poor sentence structure on sentence elements. Blame poor writers for using passive verbs badly. The notion that passive verbs are wrong is just that—a notion. Passive verbs allow us to:

- Write about events where stating an agent is not necessary.
- Summarize reports where the conclusions reached are more significant than the people who made them.
- Avoid an over-reliance on an unidentified *we* as a subject, especially in instances where the use of *we* might suggest the existence of an opposing *them.*[2]

Phrasal Verbs

A serious omission in nearly all grammar books is any mention of *phrasal verbs*. These are the verbs that consist of a verb plus a closely associated adverb, such as *abide by, bottle up, bow out, cave in, dream up, eat away, eke out, flare up, go along with,* and so on through the alphabet. I have counted almost 3,000 of them.

When the little words like *to, from, down, up* occur in front of nouns, they are *pre*positions: *over* the fence, *up* the wall, *in on* the secret. However, when they follow verbs, they are adverbs.

Phrasal verbs have the same tenses and the same forms as other verbs. Participial endings and endings that designate person, number, and tense are always appended to the verb itself, not to the adverb adjunct. Therefore, you must always write phrasal verbs as two separate words. There is no rule that you cannot end a sentence with a verb or an adverb, so it is permissible to end a sentence with a phrasal verb:

2. The debate continues about Ms. Rook's views on passive voice. See, for instance, letters to the editor in *Technical Communication,* Third Quarter 1990, pp. 221 ff, and Fourth Quarter 1991, pp. 462–463. We support her position, however, that there are good uses for the passive voice if the rules she describes are observed. *Eds.*

He didn't have the information I *asked for,* but he said he would *find out.*

Verb Tenses

Tenses indicate time. Your school grammar will point out six tenses for active English verbs. Here is a table of the six tenses using active verbs for your quick review.

	Simple	**Progressive**	**Emphatic**
Present	he calls	he is calling	he does call
Past	he called	he was calling	he did call
Future	he will call	he will be calling	
Present Perfect	he has called	he has been calling	
Past Perfect	he had called	he had been calling	
Future Perfect	he will have called	he will have been calling	

Writers have to know how to manage the tenses of verbs, that is, indications of time. Grammatical terminology does not offer much help. So let's explore each of the six tenses to understand what each tense communicates.

In standard English, the simple present tense is used to express matters that are essentially timeless or that state generally accepted conditions:

Water *freezes* at 0 degrees Celsius.
Blood *is* thicker than water.

When someone makes a general statement, "I am a test pilot," no one thinks that he or she is testing aircraft at that moment. We use the present tense not only to state occupations, but also in all matters dealing with scientific fact, established procedures, and company policy.

Verbs in the present progressive tense express present, ongoing action.

I *am balancing* my checkbook.
The company *is offering* a reserved parking space to carpool drivers.

The previous table lists three past tenses: The past, the present perfect, and the past perfect. A few minutes of reading or of listening to your friends will convince you that the simple past and the present perfect are used interchangeably when the speaker or writer is talking about a single action.

I *studied* all day.
I *have studied* all day.

When a series of events is being detailed in the same paragraph, however, the present perfect and the past perfect tenses can be used to show which event preceded another:

Millard *completed* the module testing on Tuesday. He *had been assigned* to the program in March. He *has* by now *worked* 50 hours of overtime.

The future and future perfect tenses in the previous table indicate that *will* is used to express simple future.

She *will start* to work tomorrow.

In American English *will* expresses willingness or volition; *shall* implies compulsion or a mandatory action. Its chief use is in legal documents.

Americans commonly use two future tense formations not listed in the table. We routinely indicate future action by saying, "I *am going to consult* a lawyer." To indicate compulsory actions, Americans prefer to say *is to* or *must* rather than shall:

Ward *is to (must) be* in court on Wednesday.
Helen *is to (must) adhere* to a rigid diet.

When choosing tense, we use one that is most appropriate. Some of us have unfortunately been told not to change verb tenses within a paragraph. This stricture is, of course, nonsense, and all teachers and writers should rid themselves of this common misunderstanding at once. When you are reporting on the completed status of a project—if the predominant tense in your report is the past tense, you must nevertheless use the simple present tense to state habitual and customary action and scientific results.

Mood

The word *mood* (or mode) is a grammatical term for the verb forms that indicate the attitude we have toward what we are going to say. We may be going to write something that is true or untrue, something to be done, something to be hoped for, or something doubtful. The three main categories of mood are as follows:

- *Indicative mood*—matters of fact, whether statements or questions:

  Martin *is* a good student. *Is* Martin a good student?

- *Imperative mood*—commands and requests:

  Be a good student, Martin.

- *Subjunctive mood*—matters of unreality:

  If Martin *were* a good student, he could get a scholarship.

Indicative Mood. The indicative mood consists of all the verbs we use to make statements or questions about actual events or matters of fact or fiction. All tenses of verbs are used to express indicative mood. Both active and passive verbs, verbs that take objects, and those that do not. Using indicative mood is our way of expressing anything that happens, has happened, or will happen. Verbs in indicative mood are used in objective and subjective statements and also in concrete and abstract statements. We probably use the indicative mood 99.9% of the time we write.

Imperative Mood. Verb forms in the imperative mood are the ones we normally use to give directions, make requests, and ask for advice.

Turn right at the next corner.
Pick up a six-pack on your way home.
Tell me if you think this is right.

A verb in imperative mood does not have a subject. But if it makes you happy, consider that the subject *you* is understood.

If your specification says to write procedural steps in the imperative mood, you have to use these verbs without the subject. Although we have other ways of expressing commands and requests, they are, grammatically speaking, not in the imperative mood.

The value of sentences in imperative mood is that they contrast so clearly with those in indicative mood. Whenever you use imperative mood to direct readers to perform any action, that is, anything they must set up, manipulate, operate, measure, record, or even anticipate, you can immediately follow the imperative sentence with an indicative sentence to give information. In mixing moods, there will be no confusion in readers' minds.

Subjunctive Mood. The subjunctive mood is used to express that which is unreal or that which is merely conceivable, as opposed to that which is or has been.

A clause following such verbs as *ask, insist, demand,* or *urge* also should be followed by a verb in subjunctive mood:

They asked that he *comply.*
Our boss insists that we *be* on time.
I move that the meeting *be* adjourned.

Other languages have particular verb forms to convey this mood. In English, the absence of the *s* for the third person singular and the use of *be* and *were* are all that is left of the subjunctive verb forms.

Indeed, the discussion of subjunctive mood has virtually disappeared from some grammar books, which deal with the matter as idiomatic usage. However, technical communicators sometimes need to use subjunctive mood.

Modal Auxiliaries. We use a number of helping or auxiliary verbs in English to express our attitudes about actions. Such words as *must, might, may, could, would, should,* and *ought* are called *modal auxiliaries,* because they indicate differences in feelings expressed—possibilities, potentialities, necessities, or wishes.

I *must* go.	She *ought to* go.
I *may* go.	He *should* go.
I *might* go.	They *could* go.

Technical communicators must be wary of modal auxiliaries. These words are, no doubt, what some military specifications mean when they say, "Avoid permissive language." The words *may, would, should,* and *could* allow readers a choice you may not want them to make.

MODIFYING ELEMENTS IN ENGLISH SENTENCES

Since the bare subject, verb, complement elements of sentences rarely supply intelligible information, we have a complex system of modifiers.

Modifiers of Nouns

Nouns that serve as sentence elements (subjects or complements) must often be qualified before a reader can understand the purpose of the sentence. We use articles, adjectives, verbals, and other nouns, as well as participial and prepositional phrases, to modify noun elements in simple sentences and *modifier* clauses in complex sentences.

The Articles. Grammar books devote little space to the articles, but technical writers can use *a, an,* and *the* to good advantage. If the noun is indefinite, let the reader know by using *a* or *an: a* war of words, *an* apple *a* day. If you are referring to a specific matter, use *the: the* man of *the* year, *the* winner of *the* Nobel prize, *the* seller, *the* buyer.

Also remember that articles point out nouns. In some sentences, articles are useful to readers who need to distinguish nouns from verbs. Anyone who omits the articles to save space does not have the readers' needs in mind. Crossword puzzle fans know that omitting articles from the definitions is necessary to confuse readers so that they can't tell if a needed word is a noun or verb.

Adjectives. You've no doubt been taught that adjectives qualify nouns. We speak of *gorgeous* scenery, *clever* persons, or *difficult* puzzles. You know that there are different degrees of adjectives: *good, better, best; clean, cleaner, cleanest; beautiful, more beautiful, most beautiful.* Technical communicators seldom use adjectives since few are precise or concrete.

Adjectives allow readers to supply interpretations and nuances to the meanings of these words. What, for example, is a *good* book?

Verbals. Verb forms, or *verbals*, ending in *-ing* or *-ed* are frequently used as modifiers. Such words as *moving, driving, surviving, completed,* or *submerged* add a sense of action or completion to sentences. They are often used as noun modifiers: *whistling* winds, *speeding* cars, *fulfilled* ambitions, *completed* tasks.

Nouns. Nouns frequently modify other nouns as in *concrete examples, apple trees, clay models, bird dogs, snow men,* or *American English.*

Prepositional Phrases. Prepositional phrases usually follow the noun modified as in *the apple of his eye, the book of the month,* or *plenty of resources.*

Relative Clauses. *Relative clauses* are word groups that contain verbs but do not contain the necessary elements of simple sentences. Clauses that begin with *who, which,* or *that* are called relative clauses. Such clauses can either modify closely (restrictive) or act pretty much as parenthetical remarks (nonrestrictive). For example:

> This is the house *that Jack built.* (restrictive)
> She is the one *who called.* (restrictive)
> The bird, *which you see on the fir tree,* winters in South America. (nonrestrictive)

Usage Note. For centuries writers used the relative pronouns pretty much as they chose. Fowler (1965) describes the uses of these pronouns as "an odd jumble, and plainly show that the language has not been neatly constructed by a master builder" (p. 625). His predecessors, therefore, determined to use *that* to introduce restrictive relative clauses and *which* to introduce nonrestrictive relative clauses. Fowler notes that "some there are who follow this principle now; but it would be idle to pretend that it is the practice either of most or of the best writers" (p. 626).

Fowler (p. 626) also points out that if you have chosen to use *that* to introduce restrictive relative clauses and if you have also chosen not to end sentences with prepositions, you are likely to find yourself in a dilemma. Since *that* has no objective form, you must say:

> This is the address *that* you must return it to.
> They provide a test procedure *that* the customer can rely on.

Many of us prefer the solution offered in *Webster's Ninth New Collegiate Dictionary* (1983): "*which* is used as a function word to introduce relative clauses, used in any grammatical relation except that of a possessive, used especially in references to animals, inanimate objects, or ideas" (p. 1343). Also, many of us do end sentences with prepositions, and correctly so.

Placement of Modifiers of Nouns. You can place modifiers either before or after the nouns they modify. Single-word modifiers usually precede the noun, but it is a good idea to place phrases and clauses after the noun to avoid the modifier pile-up that plagues a good deal of today's technical writing.

Articles begin nearly all sentences, followed, when present, by numbers, then adjectives, then verbals, and last, the noun modifier, as in the following example:

The three large shining metal objects . . .

Since these modifiers are all different parts of speech, no commas are required to separate them.

Modifiers of Verbs

In simple sentences, verbs are commonly modified by adverbs and prepositional phrases. Place modifiers in sentences carefully because they tend to stick to the closest word.

Adverbs. In English sentences, adverb placement is more flexible than adjective placement. Not only can adverbs precede or follow the verbs they modify, but they can also appear at the beginnings or ends of sentences.

Slowly John leafed through the pages of the brochure.
John *slowly* leafed . . .
John leafed *slowly* . . .
John leafed through the pages of the brochure *slowly.*

Most of the time you won't get into trouble placing adverbs, but you must be wary; adverbs often attach themselves to the nearest word. You must be particularly careful with certain qualifying adverbs as in this example:

Only I hit him in the eye.
I *only* hit him in the eye.
I hit *only* him in the eye.
I hit him *only* in the eye.
I hit him in the *only* eye.

Prepositional Phrases. Technical communicators use a great many prepositional phrases to modify verbs. Such phrases show time, place, manner, cause, purpose, duration, directions, or any other adverbial idea you can think of. A joke, older than most of us, is no joke, as seen in the following example:

For sale: a piano by a lady with carved legs.

This is simply an example of what happens when people do not watch where they put their prepositional phrases.

"Squinting" modifiers may occur when writers or editors move modifiers to observe a rule that is not a rule: don't split infinitives.

He planned to *carefully* read the document.
He planned *carefully* to read the document.

Will he read carefully or did he plan carefully? Careless placement of modifiers leads to misunderstanding. Splitting infinitives is not a rule of grammar; if you split an infinitive, you will not change the meaning of a sentence. Therefore, if your modifier will not fit at the end of the sentence, you had probably better split the infinitive. In the example above, you could correct the squinting modifier in one of the following ways:

He planned to carefully read the document.
He planned to read the document carefully.

SUMMARY

A knowledge of sentence structure and modifier placement (sentence syntax) allows writers to write clearly. I have therefore defined the kinds of words that can be used as subjects and objects of sentences (nouns, pronouns, verbals, infinitives, verbs, and clauses) so you will be comfortable using them in sentences.

In discussing the verb system, I have detailed the uses of active and passive verbs. I have examined the use of verb tenses according to our present needs: verbs that express general conditions (present tense) and verbs that express relative times of action (the past, the perfect, and future perfect tenses). I have also reviewed the uses of indicative, imperative, and subjunctive moods, and cautioned you about the use of the modal auxiliaries.

Finally, I have summarized the placement of the modifying elements in English sentences—those that modify nouns and those that modify verbs.

Included also are usage notes in the hope that you understand the choices open to you in matters of active and passive verbs; circumstances where split infinitives are preferable; the practical use of passive verbs; ending sentences with such adverbs as *up* and *on;* and last, the least cumbersome use of relative pronouns.

LIST OF TERMS

active voice Verb form that makes the subject the agent of the action.

clause Group of words containing a subject and a predicate.

gerund Verbal that ends in -ing and is used as a noun. Also see *verbal.*

grammar The study of a language system and the rules for forming sentences.

imperative mood Form of verbs used to give directions or to express commands or requests.

indicative mood Form of verbs used to make statements or questions about actual events or matters of fact.

modal auxiliaries "Helping" verbs that express attitudes about actions.

modifier Word or phrase that describes, limits, or makes another word or phrase more specific.

mood Verb forms that indicate how the speaker views the action.

passive voice Verb form that causes the subject of the sentence to receive the action.

phrasal verb Unit composed of a verb and a closely related adverb that is used in the same way that single-word verbs can be used.

relative clause Group of words containing a subject and a predicate introduced by *who, which,* or *that.*

subjunctive mood Form of verbs used to express that which is unreal or merely conceivable, as opposed to that which is or has been.

usage Customary or accepted practice in various speech communities or in particular technical or scientific fields.

verbal Form of a verb (particularly an *-ing* or *-ed* participle) that can be used as a noun, adjective, or adverb. Also see *gerund.*

REFERENCES

Fowler, H. W. (1965). *A dictionary of modern English usage* (2nd. ed.) New York: Oxford University Press.

Webster's ninth new collegiate dictionary. (1983). Springfield, MA: G. & C. Merriam.

SUGGESTED READING

Baron, D. E. (1982). *Grammar and good taste.* New Haven, CT: Yale University Press.

Copperud, R. H. (1980). *American usage and style: The consensus.* New York: Van Nostrand.

Evans, B. & C. (1957). *A dictionary of contemporary American usage.* New York: Random House.

Follett, W. (1966). *Modern American usage.* New York: Hill & Wang.

Gordon, K. E. (1983). *The well-tempered sentence: A punctuation handbook for the innocent, the eager, and the doomed.* New Haven, CT: Ticknor & Fields.

Hayakawa, S. I. (1940). *Language in thought and action.* (3rd ed.) New York: Harcourt.

Myers, L. M. (1968). *Guide to American English.* (4th ed.) Englewood Cliffs, NJ: Prentice Hall.

Strunk, W. & White, E. B. (1979). *The elements of style.* (3rd. ed.) New York: Macmillan.

Williams, J. M. (1989). *Style: Ten lessons in clarity and grace.* (3rd ed.) Glenview, IL: Scott Foresman.

CHAPTER 10

SWEAT THE SMALL STUFF—EDITING FOR CONSISTENCY

Vee Nelson

V. Nelson Associates

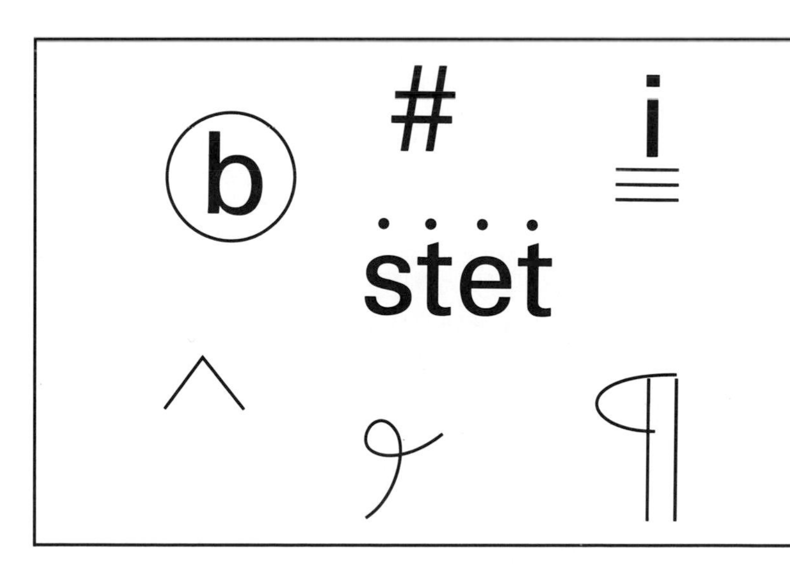

Information can be impeccably organized and brilliantly styled, yet it can still be unpolished. As readers go through a text, if they see "percent" on page 3 and "per cent" on page 5, "chairman" on page 2 and "chairperson" on page 8, these annoying inconsistencies will stick to them like burrs.

Such inconsistencies may seem like small stuff to most writers; after all, they don't take away from the *meaning* of the text. Yet, small as they seem, just these sorts of inconsistencies may, in fact, hinder readers' acceptance of the writer's message. Not only can inconsistencies annoy, but they can also lead readers to question the credibility of the writer and publisher.

Most of the time, it is left up to an editor to "sweat the small stuff": identify and correct the inconsistencies of wording, punctuation, or style. The professional editor is trained to see inconsistencies and to make choices between two or more equally correct forms. With the help of dictionaries, style manuals, precedents, and astuteness, for instance, the editor determines that "percent" is to be one word and "chairperson" best reflects the company's position on nonsexist language.

Admittedly, writing and editing are two distinct tasks, but all writers can gain an edge by being aware of the small stuff and ways to control it. With this awareness writers not only make the editor's job easier but also are able, if called upon, to take up the editor's blue pencil and bring editorial consistency to the text.

This chapter covers information and techniques that can help you achieve consistency. Specifically, in this chapter you learn how to do the following:

- Recognize elements that have two or more correct forms and that need editorial consistency.
- Use a dictionary or style manual to gain consistency.
- Develop a style sheet to manage copy style choices.
- Use the style sheet to facilitate team or recurring writing projects.

WHAT ARE VARIANTS AND HOW DO THEY AFFECT EDITORIAL CONSISTENCY?

In editorial workshops, I can tell the difference between writers and editors with one question: "What needs to be consistent in a manuscript?" The writers respond with "verb tense," "point of view," or "tone." The editors nod in agreement, but add "spelling," "hyphenation," "capitalization," "format."

Actually, both types of consistency are necessary. Verb tense, point of view, tone, all of the concerns outlined by writers, are essential to well-presented information. Yet, they carry the manuscript only so far toward being a polished product. Without the editor's concern for the consistency of *copy style,* as I will call it here, the manuscript looks sloppy and unreliable.

The sentence below contains no "errors," but there are many issues where authorial or editorial preferences could bring about change.

> We must designate one person to be the decision maker; he will be responsible for selecting the data-base manager, word-processing package and a graphics program for our department.

One of these issues pertains to writing style: whether to connect clauses with a semicolon or a period. The other issues pertain to copy style:

- Decision maker, decision-maker, or decisionmaker
- He, she, he or she (*not,* he/she or s/he, I hope)
- Database, data-base, or data base
- Word processor, wordprocessor, or word-processor
- A comma before *and* or no comma

Copy style encompasses all of the conventions that determine which *variants* are used in a manuscript. A variant exists when there are two or more equally acceptable forms of the same element. Variant spelling is the most obvious. As shown above, there are three ways to spell *decision maker, data base,* and *word processor.* All spellings are correct, and it doesn't matter which form you use, as long as you use it consistently throughout a given manuscript.

Spelling is only one aspect of language that has variant forms. You can also find variants in capitalization, abbreviations, numbers, and even punctuation and grammar. Here is a list of the areas in which you may find variants:

spelling	units of measure
compound words	quotations
hyphenation	illustrations and graphics
capitalization	captions, legends
abbreviations	tables and graphs
numbers	documentation, reference
math symbols and signs	bibliographic form
titles	dates
honorifics	foreign terms
names	emphasis
short names	list format
cross references	special terms

To these can be added some points of grammar and punctuation, as well as some issues of format, such as headings, margin width, paragraph indentions, or margin justification. I will limit the rest of the discussion primarily to consistency issues concerning mechanics, spelling, and language.

Your immediate reaction may be to say, "But there are rules for most of these." Yes, there are some rules, but not as many as you were, perhaps, taught in school. Look at spelling again. Most words have a fixed spelling, but the English language also has thousands of words with variant spellings. Most have two possible spellings, such as *judgment* and *judgement*. Others can have several correct possibilities: *goodbye, good-bye, good bye, good-by,* and *good by.*

Similar lack of consensus occurs with capitalization. Almost all of the time, it is "correct" to begin a sentence with a capital letter. What, however, about short sentences in lists? Sometimes the initial word is capitalized; sometimes it isn't. The "rules" vary from text to text, and arguments abound among editors as to the "correct" style for capitalization and other seemingly straightforward issues.

Editors must decide issues of consistency, too, with some points of grammar. For instance, is *data* allowed as singular? Is the past tense of *dive* going to be *dived* or *dove?* There is no right or wrong to these choices—as long as the choice is made consistently.

You need to make a decision about consistency any time there are two or more correct forms of the same word or element. For instance, you can write *wirefeed* or *wire feed* and be correct. Or you can write the date as *September 4, 1992; 4 September 1992; Sept. 4, 1992; 4 SEPT 92; 9/4/92; 9-4-92;* and several other ways. All are correct, but you must select one form and stay with it throughout the text.

HOW TO USE A STANDARD DICTIONARY AND STYLE MANUAL TO ENSURE CONSISTENCY

The starting point for ensuring consistency is to select a dictionary and style manual that will be the standard; that is, they will serve as a base for your copy-style decisions. You then consult these references for decisions on how to handle variant forms.

Dictionaries

Dictionaries are essential for helping you determine spelling, plural forms, and proper names. You may consult several dictionaries to help you select the right word for the meaning, but you should select only one dictionary and stick with it to manage variant forms.

Although most of the dictionaries published in the United States use the same first and second choices for spelling, there are some discrepancies. Compare entries of three current dictionaries—*The American Heritage, The Random House College Dictionary,* and *Webster's Ninth New*

Collegiate Dictionary. Two of them cite *TelePrompTer* as the trademark; one cites *Teleprompter.* Two agree on *good-by* as the first spelling; one lists *good-bye* first. Two have the adjective *long-time* and the past tense verb *programmed;* one has *longtime* and *programed.* Elsewhere, two have the noun *air freight;* the third has *airfreight.*

A good dictionary is essential for creating consistency, but the same dictionary has to be used by everyone who marks the manuscript. The best policy is to designate a "house" dictionary and use it as the starting point for all copy-style decisions.

Style Manuals

Some words are not listed in the dictionary. For example, none of the three dictionaries just mentioned tells you whether to use *decision maker, decision-maker,* or *decisionmaker.* None includes *filename* (*file name?*) or *lite.* In these instances, you need another tool—the *style manual.*

Style manuals are essential for helping you maintain a consistent copy style. Sometimes companies have an approved style manual, which has either been developed in-house or bought from a publisher. Other times, you may have to select or develop your own. In any case, it is important to have this reference so that you can know what the preferences are and follow them as closely as possible. For instance, if you are writing a training manual for a software program, you would consult a style manual to determine whether to use *database, data-base,* or *data base* or whether keys should be in all caps (ENTER), boldface (**Enter**) or in another type face. It helps to know whether you should use bullets (■) or some other indicator for lists.

Depending upon your field, you can choose from several style manuals. Primarily, you look for a manual that covers the types of questions that you face most often. At the end of this chapter is a list of some of the major style manuals you may want to use.

WHEN TO OVERRULE SUGGESTIONS IN DICTIONARIES AND STYLE MANUALS

Even if everyone agrees to use the same dictionary and style manual, writers and editors can go only so far in managing all of the variant forms that might appear in a manuscript. Authorial, editorial, or managerial preference may require that a different style choice be made.

Although most style manuals say *not* to capitalize titles except when they precede the person's name, in your company it may be expected that you capitalize President, regardless of where it appears. Similarly, you may decide that department names should be capitalized, although your style manual says not to.

You may also find that style manuals don't always address the diversity that often appears in technical information. Although the style manual may say to hyphenate unit modifiers before a noun, the hyphenation may create an awkward combination, given the lengthy technical phrases you are using. For instance, if you follow the "rule" of hyphenating unit modifiers, you could wind up with a string like this one: *multi-module-minimum-cavity-low-profile-hot-glue gun*. But such a string looks awkward, and you probably won't find a style manual to help you decide how to handle this situation. So you must make choices to suit the situation and do your best to note these *ad hoc* style choices.

All of the possible variations can overwhelm even the most experienced writer. When the writer is working with a team or when one person is wearing several hats, managing all copy-style issues can become even more difficult. What you need is a system to help you keep track of all of your choices.

WHAT A STYLE SHEET IS AND HOW TO DEVELOP ONE

A *style sheet* is a means of tracking the copy-style choices you make as you read through a manuscript. You can create a style sheet in a number of ways, but I will describe two simple methods: the box format and the checklist format. In the box format, you draw a single, centered vertical line and two equally spaced horizontal ones intersecting the vertical line to create six boxes, as shown in Figures 10.1 and 10.2. In the checklist format, you make a list of the variants you want to track, and under each topic, you write down the acceptable choice, as shown in Figure 10.3.

I find it easier to begin with the box format. If I will be using these copy-style choices later in other documents, I turn the items in the boxes into a checklist. This is the sequence I will discuss here. Before you begin editing, you'll want to create several pages of boxes. Once you have been through the process using this approach, you may want to modify it to suit your situation.

1. **Scan the manuscript.** Even if you wrote the manuscript, don't try to remember what possible variants you have. Before you start the editing process, take time to scan the document to determine how many possible variants you have.
2. **Distribute the letters of the alphabet among sections of a page of your style sheet.** On a page of your style sheet, list the alphabet so that the letters are divided evenly among the blocks (see Figure 10.1). You may use one or two pages for

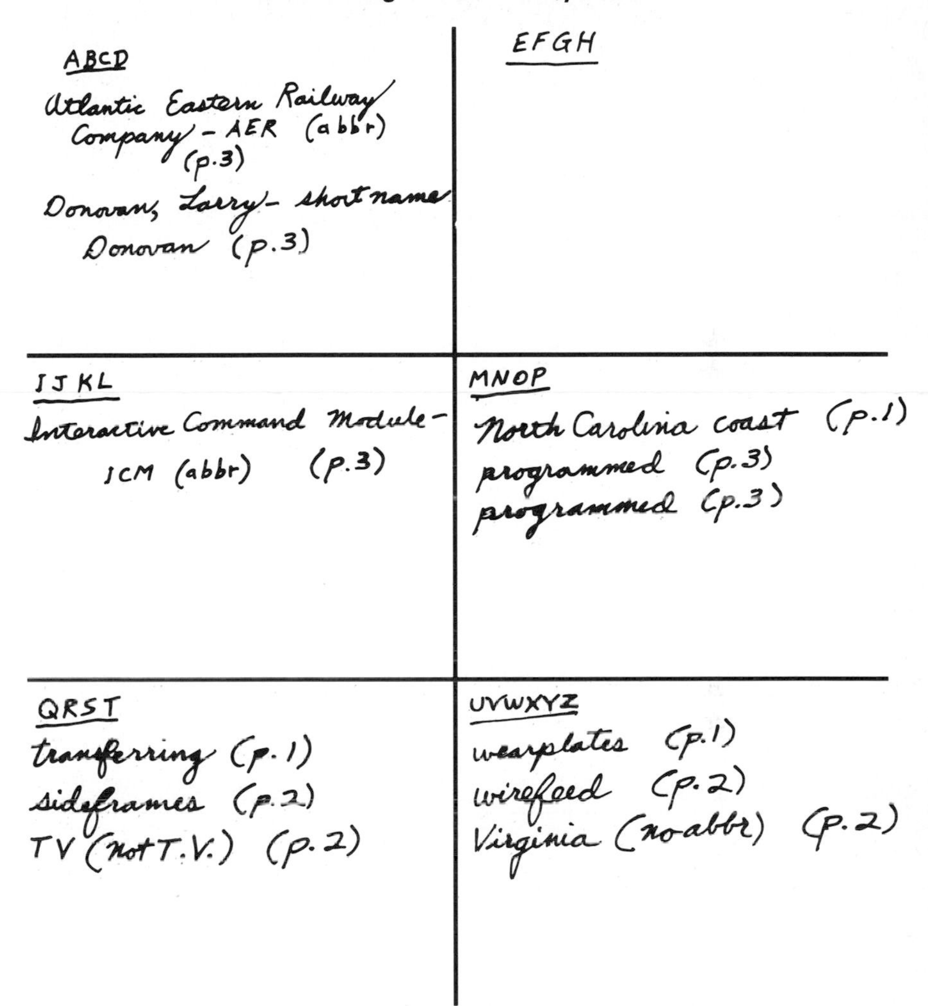

AER Welding—Checklist Style Sheet
ABCD
Atlantic Eastern Railway Company—AER (abbr) (p.3)
Donovan, Larry—short name Donovan (p.3)
EFGH
IJKL
Interactive Command Module—ICM (abbr) (p.3)
MNOP
North Carolina coast (p.1)
programmed (p.3)
programmed (p.3)
QRST
transferring (p.1)
sideframes (p.2)
TV (not T.V.) (p.2)
UVWXYZ
wearplates (p.1)
wirefeed (p.2)
Virginia (no abbr) (p.2)

FIGURE 10.1 Example of Alphabetized Variants in a Style Sheet Using the Box Format

the letters, depending on how many spelling variants you found as you scanned the manuscript.

3. **Select the categories of variants to include in the style sheet. Label the sections on another page of the style sheet.** Decide which areas have the most variants. If, in scanning, you notice that you used a number of abbreviations or that you have numbers on almost every page, then you should have "abbreviations" and "numbers" as category headings. Put one category in each section of the style sheet (see Figure 10.2).
4. **List selected forms under the appropriate headings as you edit the document.** As you go through the editing process, record your

AER Welding—Checklist Style Sheet

CROSS-REFERENCE	HYPHENS
(Figure 1) – *not* (See Figure 1) (p. 1)	number–adjective combination eg. 36-year (p. 1)
NUMBERS	**MEASUREMENTS**
spell out under 10 (p1)	use figures with – e.g. 20 ft. (p. 2) use symbols for – e.g. 3/8" (p. 2)
PUNCTUATION	**GRAMMAR**
no comma after short introductory phrase e.g. In 1979 we had (p2) dashes with space on either side (p. 2)	Pronoun reference – *it* for Companies (not *they*) (p. 2)

FIGURE 10.2 Example of Categories in a Style Sheet Using the Box Format

selected form under the appropriate variant heading. For instance, if you have decided to precede each item in a list with a bullet, write "bullet before each item" under the variant heading "Lists." It is also best to put the page number where the first instance occurred; that way, if you want to go back and change it (for reasons to be discussed), you can find it easily.

5. **Edit the style sheet.** As you continue editing, you may find that you aren't happy with your first choice for handling a variant. In this case, scratch out the original choice and replace it with the new one. If, for example, further into the manuscript you realize that you have used mostly *database* instead of *data base,* then you will save time by changing your style sheet instead of changing

AER Welding—Checklist Style Sheet

Compounds

sideframes
wearplates
wirefeed

Cross-references

(Figure 1)
NOT (See Figure 1)

Names and Abbreviations

Atlantic Eastern Railway Company
short name—AER, NOT the Railroad, NOT Company or AE
pronoun ref.—it, not they
North Carolina coast
Larry Donovan
short name—Donovan
Virginia (no abbreviations)
Interactive Welding Model
abbreviation IWM, after first reference

Numbers and Measurements

Generally under 10, spell out
e.g., three choices
All figures with measurements, time, ratings, cross-references
e.g., 20 feet, 8 minutes
Measurements expressed in symbols
e.g., 3/8″
NOT 3/8 inches

Variant Spelling

programmed/programming
transferring
TV, not T.V.

FIGURE 10.3 Example of a Style Sheet in a Checklist Format

all of the one-word spellings to two words. Your first draft of the style sheet will inevitably have many changes. Each time you make a change, however, be sure you record the page number. When you have completed the manuscript, you can go back and edit your list, especially if the style sheet is going to be used by anyone else or placed in a file for future reference.

The important thing to remember when developing the style sheet is that you include *only* the variants. You need not record something like *receive,* even if it is a difficult word for you to spell, because any dictionary you look in will spell the word that way. The goal of your style sheet is to remind you of choices made, so list only words like *data base* or others for which you selected a variant in spelling or other copy-style choices. Likewise, do not include "period at the end of a sentence" or "capitalize the first person singular pronoun," because these rules never vary. Instead, record "comma before a conjunction in a series" or "capitalize the first word in each item of a list" because these are not absolute rules.

HOW TO USE THE STYLE SHEET IN TEAMWORK AND RECURRING PROJECTS

After you have finished editing the manuscript, don't throw away the style sheet. You can use it in the publication process, for improved teamwork, and for recurring writing projects.

To expand the usefulness of the style sheet, you first need to turn it into a more easily accessible form (and probably a neater one, too). The best format is a checklist, which you can create in the following way:

1. Look back at the style sheet you have and determine the major categories that you wound up with. (You may have discovered that some initial categories can be combined with others and some can be deleted or added.)
2. Write these categories in alphabetical order and list the *preferred* variants under the appropriate headings. It is best if you also alphabetize the list. Your checklist will look something like the one in Figure 10.3.
3. Keep the checklist on paper or put it on disk. The latter can be useful for updating the checklist easily and making available an online checklist.

Whether on paper or disk, the checklist can be useful in several ways.

- **To smooth out the publication process.** You can pass a copy of the checklist with your document to a typist, an editor, a proofreader, or anyone else in the publication process. Your typists will be able to improve the consistency of their work if they know what you intended to be the pattern. Editors, if you have any, might make some changes, but at least they will know what choices you have made and can therefore make more efficient decisions about the final style for the document. As the document goes through the publication process, the checklist should go along. In this way, everyone at every stage knows what variants have been selected and can contribute to making the document consistent.
- **To encourage teamwork.** If you are involved in a project with a team of writers, the style sheet and checklist are essential. Before starting

the project, someone should be appointed "editor" (if there is not already one). That person should designate the standard dictionary and style manual, and, with the agreement of all members of the team, establish choices for all obvious variants that might arise. If you are working on a software manual, for example, you automatically know that some terms, such as *filename,* and some mechanics, such as handling names of keys, are going to become consistency issues.

Each team member can then keep a style sheet on the sections he or she is writing. At various checkpoints, whenever the team meets to discuss the project, one agenda item should be the style sheet. Members reach some agreement as to the variants used, settle questions that have not yet been answered, and modify or add to the individual style sheets. Before the final drafts are written, the editor clears up any discrepancies and compiles the style sheets into a checklist, which everyone then uses to make final changes in each member's portion of the document.

- **To ensure consistency in recurring projects.** Whether you are working on a team or individually, you can use the style sheet for recurring projects. By keeping either printed or electronic copies of the final, revised checklist, you are starting your own style manual. Freelancers who work for numerous clients find this technique useful. They create an online checklist with a file name that represents the client's name, for instance, "style1.ibm" (DOS version) or IBM style (Macintosh version). Every time they do a project for IBM, they either build on that style sheet or create another one with the same extension, such as "style2.ibm" (or IBMstyle2). Each time they do a subsequent project, they know that they are using the same variants as before. This makes their writing job easier and also presents a better, more consistent product for the client.

Using style sheets as a basis for consistent copy style also works well for in-house communicators. By creating a base for your style sheet and then adding to it each time another project comes along, you eventually build an in-house style manual. The more you add, the more complete it becomes and the less time you have to spend in the future making decisions about style. The style sheet, which can be either on disk or on paper, can lead eventually to a more formal, in-house style manual.

SUMMARY

This chapter explained the importance of consistency in writing and copy style. It explored variants, those elements in the language that have two or more correct forms, and explained how to select a standard dictionary to resolve most matters of spelling and a style manual to resolve most matters of mechanics and formatting.

Because dictionaries do not resolve all inconsistencies in spelling and style manuals do not resolve all matters of mechanics or formatting, this chapter explained why and how to develop a style sheet. You develop a style sheet, which records ways of handling variants, by doing the following:

1. Scanning the manuscript and identifying possible variants.
2. Selecting categories of variants to include in the style sheet and labeling those sections.
3. Distributing the letters of the alphabet among sections of another page of the style sheet.
4. Listing your choices under the appropriate variant categories.
5. Editing the style sheet.

In addition to resolving inconsistencies in a single manuscript, style sheets can be used to smooth out the publication process, encourage teamwork, and ensure consistency in recurring projects.

Whether you are a writer or an editor, your attention to matters of consistency can have several benefits for you. Paying attention to consistency and working with the established editorial style can reduce the work that others may have to do to publish or produce a document. Attention to matters of consistency can also give you a new awareness of ways to make your documents more polished.

Whether you are writing an article, a newsletter, a brochure, or a training manual, the product won't look polished and professional until it has a consistent copy style. This consistency is not solely the responsibility of an editor. Today, especially with increasingly widespread use of desktop publishing, all communicators have responsibility for the final product. Even with the help of an editor, communicators find they can ease the editing process by making the document as consistent as possible before passing it on to the editor. Either way, communicators find it is to their advantage to "sweat the small stuff."

LIST OF TERMS

copy style Spelling, abbreviations, capitalization, italics, numbers, punctuation, and all other elements of usage that have two or more equally acceptable forms. Consistency is the key word for a good copy style.

style manual Set of the preferred conventions to be followed to create a consistent copy style. May be either published and widely used by many professions or created in-house for use in a particular department or with a particular kind of document.

style sheet Listing of copy-style forms to be used in a given manuscript. Created during the editorial stage of manuscript preparation, it may supplement a style manual or be used independently.

variant One of two or more acceptable forms for the same word, usually associated with spelling, but may also refer to any aspect of copy style.

SUGGESTED READING

American Psychological Association. (1983). *Publication manual of the American Psychological Association.* (3rd ed.) Washington, DC.

This manual is specific to the psychology and sociology fields. It gives the standard copy style and formats for submitting articles in these fields. Some editors find it is difficult to use and not general enough for editorial work in other disciplines.

The Associated Press stylebook and libel manual (rev. ed). (1987). Reading, MA: Addison-Wesley.

Arranged alphabetically and easy to access. Entries on spelling preferences and trademarks are the best. Others, such as those on grammar and capitalization, are too brief. Primarily used by most newspapers, public relations firms, and advertising. It can also be useful for people who do newsletters.

CBE style manual (5th ed.). (1983). Bethesda, MD: Council of Biology Editors.

The first part is general and suitable for any kind of editing and publication work; the last part is heavily scientific.

The Chicago manual of style: For authors, editors, and copywriters (13th ed.). (1982). Chicago: University of Chicago Press.

The most prestigious style manual, and probably the most widely used. It is heavy on scientific fields, light on general areas. It is particularly helpful for issues such as tables and references.

McGraw-Hill style manual. (1982). New York: McGraw-Hill.

This is another general style manual, similar to Skillin and Gay's *Words into Type,* but also includes details on math or scientific elements. This latter information is useful for some people who work with technical information.

Skillin, Marjorie E. & Gay, Robert M. (Eds.). (1974). *Words into type* (3rd ed.). Englewood Cliffs, NJ: Prentice-Hall.

This is a useful style guide for general use. It includes not only major issues in publication procedure and copy editing style, but also an excellent section on grammar, word usage, and punctuation. Some parts, such as typographical style, will also be useful to people working in desktop publishing.

U.S. Government Printing Office style manual. (1984). Washington, DC: GPO.

The grand-daddy of style manuals. It was the first one published in the United States, and, as a result, it retains some outdated style conventions. Yet it is excellent for compounds, hyphens, and similar issues of copy style. It is also essential for people working with U.S. government documents.

Webster's standard American style manual. (1985). Springfield, MA: Merriam-Webster.

Far less technical on the composition, design, and production than other style manuals named, but includes all other elements of copy editing and is fairly easy to use.

PART IV

TECHNIQUES FOR VERIFYING AND PROTECTING INFORMATION

CHAPTER 11

TEST DRIVE—TECHNIQUES FOR EVALUATING THE USABILITY OF DOCUMENTS

Ann Hill Duin

University of Minnesota

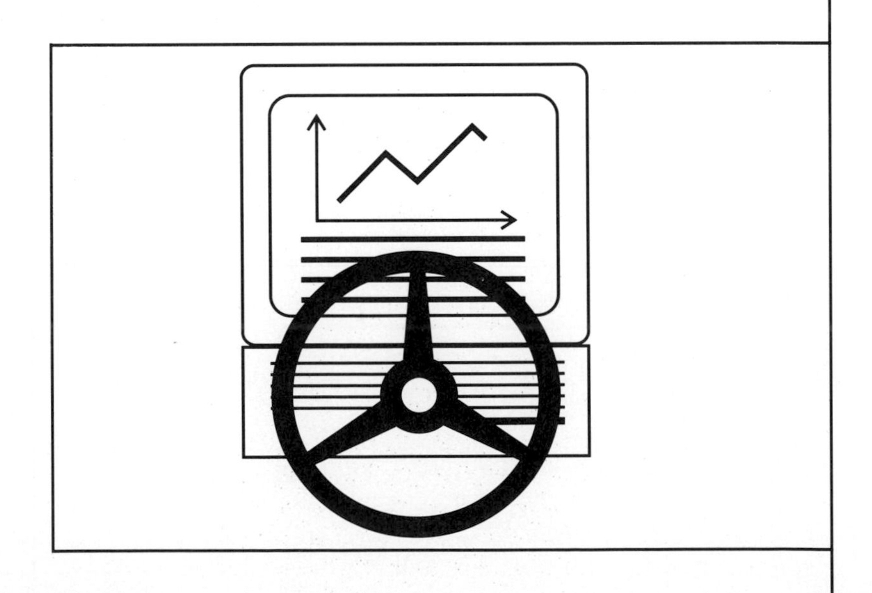

Consider these problems with products and their information:

- Candle Corporation's main product is a computer performance monitor for large IBM mainframe computers. In the past Candle's users were a small, highly technical group that readily understood the product's 800 commands—each with its four-letter acronym. In the last few years, however, the audience has grown and changed considerably. Users are no longer exclusively technical and could not be expected to know all of the commands. Product developers saw menus as a way to make the product easier to use for the less technical users. While the menus were initially well-received by the sales force and new customers, the menus were not designed according to the users' tasks, and, as a result, they were not used (reported in Tavlin, 1989).
- The Illinois Institute of Technology needed an in-house manual for its accounting system. The information successfully went through many usability tests with users. Yet after the manual was distributed, the cashiers in the bursar's office rarely—if ever—used the information (reported in Mirel, 1987).
- The Department of Defense has maintained a major research program in document design for the last 35 years. Their goal has been to develop task-oriented manuals for novice users. The primary approach has been to develop guidelines and specifications for writers to use when developing information, and yet, manuals continue to be inadequate. In one case, a radar technician had to refer to 165 pages in eight documents and look in 41 different places to repair one malfunction (reported in Duffy, Post, & Smith, 1987).

This chapter explores how these and similar problems can be resolved through usability testing. Specifically, I address these questions:

- What is usability?
- How do you test usability?
- How do you design, plan, and conduct usability tests?
- How do you report the data collected in a usability test?

WHAT IS USABILITY?

Whether a device being described is an automatic teller machine, a glucose monitoring device, a sophisticated coffee maker, or a computer, users demand information about the device that is tailored to their goals and centered around their needs. To do this, technical communicators conduct usability tests. The tests let technical communicators

- Discover users' goals and needs.
- Plan for users' needs.
- Design documents with users' needs in mind.

That is, technical communicators must "test drive" information to make sure people can understand it and make effective use of it.

A coffee maker, computer, or car is usable only if specific users can achieve specific goals efficiently and effectively within a particular environment. That is, *usability* is the degree to which an intended audience can perform the desired tasks where those tasks are usually performed. Booth (1989) notes that usability problems afflict all types of complicated products, from intelligent tutoring systems to everyday household items. He states that:

> The issue of concern is how to mitigate the effects of these usability difficulties, or better still, how to ensure that usability problems never arise. This challenge is best expressed in the statement: Today we are just as capable of producing an unusable product or system as we have always been. In other words, the challenge is this: although we might recognize usability as a central issue in the design of complex products, how can we ensure that future products do not suffer from these problems? (p. 104)

While technical communicators can't ensure that all future products and information will not suffer from usability problems, technical communicators can alleviate many of these problems by testing the usability of products and information before bringing them to the public. Testing the usability of information as part of the product's design and development process can improve the quality of the product as well as the accompanying information.

HOW DO YOU TEST USABILITY?

Usability testing is taking a product and its information for a "test drive." In other words, usability testing is a procedure in which people who are representative of the intended users of a product are asked to use the product and its information before the product is released to the public. In this way, problems identified in the test can be addressed before the product is released. This section explains where you can perform usability tests, what you should look for in usability tests, and why you should conduct usability tests.

Where You Can Perform Usability Tests

You can perform usability testing either in a controlled situation such as a laboratory or in the user's own environment, which is usually called the field.

In usability tests conducted in a laboratory, you construct an environment like that of the intended users and watch them use the product and its information. Because the environment is controlled, you can see the initial causes of users' problems and how users respond to them.

Laboratory tests are usually viewed from behind a one-way window so the testers do not interfere with the test. These tests are often videotaped and audiotaped, so product designers and technical communicators can have an extensive record of data to study later. For example, testers can construct a home or office setting with either hidden or visible recording devices, and then testers can ask users to use the systems or complete specific tasks. In one case, such testing identified a critical issue when the testers learned that an initial step for installing a hard drive appeared in a computer file that users could access only after they had installed the hard drive.

In contrast to laboratory tests, usability tests conducted in the field involve "investigations done in the natural environments of those under study" (Gould & Doheny-Farina, 1988, p. 330). By studying how documentation is used in a natural work environment, technical communicators can learn much from only a few users. Thus, in the field, testers usually watch people use a prototype of the product in a real-world setting. For example, testers can watch people program a sound system or they can identify and study those people who use an electronic mail system. *Field tests* show firsthand who a product's users are, what they do, and how they do it (Brooks, 1991).

What You Should Look for in Usability Tests

In testing, technical communicators study the users' degree of effectiveness, efficiency, and comfort when using the product and its information. Technical communicators use the data gathered to evaluate the usability of a product and its information, so testing is also called evaluation. Technical communicators can perform two types of testing or evaluation: summative or formative.

Summative Evaluation. *Summative evaluation* involves assessing the impact, usability, and effectiveness of the product and its accompanying information. Summative evaluation mainly includes quantitative measures and checks for verification. The product and its information are tested at the end against preset criteria, and they either pass or fail. For example, a VCR installation guide might be tested to determine whether users can install the VCR in 20 minutes or less. If the product and its information fail to meet the criteria, technical communicators review the videotapes, audiotapes, observation notes, and other data, as well as interview test subjects, to determine the cause of the problem.

Another form of summative evaluation involves testing the product at the beginning and the end of the design process and then measuring the differences between the two tests. For example, an early prototype of installation instructions might be tested to see if the most important steps are included. Once revisions are made, the product and the accompanying information can be tested again to see if the revisions resulted in less time needed for users to install the system.

Formative Evaluation. *Formative evaluation* helps product designers and technical communicators to evaluate the product or information in its formative stages and apply the results as they redesign and refine it. Formative evaluation requires quantitative and qualitative information to help designers pinpoint parts of the product that require alterations and to help communicators pinpoint documentation problems. Formative evaluation is a form of diagnostic testing in which testers look at the product and its accompanying information, see how well people are performing tasks, find weak spots, improve the product and information, and then repeat these steps. For example, an early draft of a tutorial for a program that allows users to write checks might be tested to see which tasks users most want to perform and how long it takes users to set up the program and complete these tasks. In this way, testers can locate weak spots—perhaps some users don't want to create check forms but simply want to maintain a check register—and then revise the information for different levels of users.

Although formative evaluation lends itself to laboratory testing and summative evaluation lends itself to field testing, both forms of evaluation can be conducted in either setting.

Why You Should Conduct Usability Tests

You should conduct usability tests to make information usable, improve products, remove obstacles, and save costs. Grice (1991) states that:

> We are surrounded by so much information today that the real challenge to technical communicators is not so much to produce information as it is to make information usable. If we are to succeed in making information usable, we must consider usability an inherent part of the process we use to develop information. Thus, testing information for usability is an important part of a technical communicator's job; it can even be a career in itself. (p. 178)

According to Dumas (1989, p. 37), we should conduct usability tests to achieve the following objectives:

- Raise the awareness of product designers about usability issues.
- Improve the technical and managerial skills of product designers and product managers.
- Stimulate cooperation among design team members, such as engineers, writers, and human factors specialists.

By performing usability tests, technical communicators not only can help improve a product and its information, but also can affect the future design and development of products within their organizations.

According to Redish (1989), we should conduct usability tests to identify obstacles before a product is shipped. Ultimately, removing these obstacles increases sales, reduces the need for updates, reduces complaints, saves the cost of help calls and service calls, increases the likelihood that customers will buy another product from the company, and saves future embarrassment.

But the need for effective documentation is not just a consumer issue. The quality of documentation has significant cost consequences in the use of the technology. Fifteen years ago Shriver and Hart (1975) researched data on the speed and accuracy of job tasks performed by individuals using information written according to two different specifications. The researchers estimated that the publisher of the information could save $830 million through reduced maintenance costs and another $900 million in reduced training time by selecting the more usable specifications.

HOW DO YOU DESIGN, PLAN, AND CONDUCT USABILITY TESTS?

When you begin a test, you need to decide the following:

1. The types of things you would like to learn about a product and its information (i.e., your test objectives).
2. Whether to conduct the test in a laboratory or in the field.
3. Specifics of testing:
   - Particular test(s) you'll conduct
   - Who the subjects will be
   - When the test will be conducted
   - Who will be on the test team
   - What procedures you'll follow
   - What you'll do with the data throughout the design, planning, and testing process

The most critical criteria are that you have subjects who represent the audiences, that you have the subjects do realistic tasks with the product and the information, and that you watch and learn from the subjects (Redish, 1989). The next several sections explore how to design, plan, and conduct tests.

Determining Test Objectives

Your first task in designing a usability test is to determine what you need to learn about a product and its information; this involves determining your test objectives. Wright (in press) states that "it is not appropriate to think of the objective of evaluating a text as being akin to 'seeing if it works' or 'seeing if readers understand it'" (p. 324). Different test objectives will result in the detection of different types of problems.

Wright uses the analogy of various types of filters when thinking about the objectives for a usability test. If your objective is to optimize the documentation, then you need to catch all the lumps, no matter how small. In this case the mesh needs to be very dense. On the other hand, if your objective is to avoid catastrophes within the documentation, then the mesh need only be so dense as to remove the very large lumps.[1]

Queipo (1991) states that objectives are simply goals that you want to meet, noting that "a good test objective should paint a picture in the user's mind; it should allow the user to visualize what is being looked for" (p. 186). Queipo offers this example: if you are testing the usability of a machine, the test objective might be that the user should understand when to change the mode of the machine. Then, as the tester, you would need to list exactly what the user is expected to do when performing the task and what must be attempted by the user before the task is complete.

Thus, good test objectives describe which actions to expect and tell testers how to distinguish between acceptable and unacceptable performance. Test objectives, then, are similar in format to objectives for documents, which are described in Chapter 2. The primary goal is that each test objective indicate a level of performance users should be able to achieve. Consider, for example, Queipo's example for test objectives for a user's guide:

- The evaluator [user] must locate the proper section in the manual within one minute.
- After locating the proper section, the evaluator [user] must locate the LOGON command within 30 seconds.
- No fewer than 90% of the evaluators [users] must be able to perform objectives 1 and 2 within the specified times.
- No fewer than 95% of the evaluators [users] who locate the LOGON command must be able to demonstrate correct use of it two weeks after the test session (p. 187).

These test objectives appear designed to "catch all the lumps, no matter how small." Keep in mind that alternative test objectives might be to provide input for later document design decisions, to contrast two alternative designs for the documentation, or to develop questions to include on forms inserted with the product and its information (Wright, in press). In any case, your test goals should reflect what you want to learn about your product and its information. For example, if you were testing the map or route system for users to locate the emergency room in a hospital, your goal is to avoid mistakes within the mapping system, and your objective might state that all users should locate the emergency

---

1. Wright goes on to caution that the filter analogy, as with most analogies, has its limitations, for it can be misleading to think of usability testing as a filter; if problems are not detected, this does not mean that problems are not present.

room within a specified time after they enter the hospital. Or, if you are initially testing two types of help systems to accompany a software program, your goal might be to learn which system users access more often, and your objective might state that users should be able to access and use the help system from any point in the program.

Deciding Whether to Conduct a Test in a Laboratory or in the Field

Your next task in preparing a usability test is to decide whether to conduct it in a laboratory or in the field. These alternative settings affect the type and quality of the information you acquire.

Laboratory Tests. The major emphasis in laboratory tests is on systematically observing people using the product and the documentation (Sullivan, 1988), although methods vary according to test objectives. According to Dumas (1989), with formative evaluations conducted in a laboratory setting, you can achieve the following objectives:

Type of Test	Description
Have designers and users work together with a product.	Performed during early design stages to diagnose major problems with a new design. For example, designers and users could load networking software for a new telecommunications system, discussing problems in both the design of the product and the format of the documentation.
Test pairs of users who work together to accomplish tasks.	Results in a rich dialogue between the users through which users discover insights from each other concerning how to use the product, and designers can learn how to redesign based on these insights. Can also be used to identify how users collaborate to construct a product.
Have a single user attempt a task alone.	Simulates the situation in which a single individual uses a new product without assistance. Can be used to indentify which parts of a tutorial individuals will use when left on their own and what strategies they follow when something doesn't work or they get lost.

Laboratory tests typically have between five to fifteen test subjects; however, the paucity of subjects is compensated by the richness of data collected. Laboratory tests let technical communicators know exactly what is being studied and how the user and the product are interacting (Redish & Schell, 1989).

Field Tests. Gould and Doheny-Farina (1988) state that although field tests don't represent controlled studies such as those that can be conducted in the laboratory, technical communicators can learn the following:

- Who users are (company demographics and personal information)
- What users' work environments are like
- Which users have access to what information
- How information is used in those environments
- Why users behave the ways they do (the meanings users attribute to their behavior)
- How accurate and useful the documentation is
- How satisfied users are with the documentation

The Digital Equipment Corporation recently decided to change its focus from laboratory tests to field tests for two reasons:

- Although laboratory tests often met company goals, Digital found that these goals were usually not grounded in customer experience.
- Digital became aware of recent theoretical and philosophical work (e.g., Winograd & Flores, 1986) that argues for the importance of context in understanding human behavior (Good, 1989).

Digital's primary technique for designing and building usable systems is to visit customers and collect data at the user's workplace. Engineers interview users actually working with their computer systems, ask about users' work, the details of system interfaces, and how users perceive the various components of the systems. In this approach to field testing, users and engineers work together to learn how users actually use systems. Engineers also interview users after they have used a system, learning how users feel about a system after they have used it. Engineers use the data collected to build a theory of usability and follow this theory throughout the design process. In short, engineers analyze the data about users' experiences to provide detailed information for designing specific products and to generate ideas for new products (Good, 1989).

It is the rich influence of context (real-world situations) that has led organizations to study the usability of information in the field. Users in different contexts have different usability needs within the context of the workplace. For example, a person's use of an electronic mail system for collaboration with local colleagues may differ from the use of the same system for collaborating with people nationally or internationally. Wright and Monk (1989) state that:

> It is useful to think about usability problems as forming a hierarchy with each successive level in the hierarchy requiring more contextualized knowledge for its diagnosis . . . An example of a problem at the bottom of the hierarchy is a user having difficulty in correcting typographical errors because of inadequate delete facilities . . . An example of a problem near the top of the hierarchy is the difficulty that a user might have in understanding how to carry out a database search task on some new system. The problem arises from the user's experience with the system, with other systems and with the task it is being used for and so depends very much on the context. (p. 26)

How to Plan a Usability Test

Effective writers report that 80% of their writing time involves planning. For them, once a good plan is in place, the drafting, revising, and editing take only 20% of the writing time (Selzer, 1983). In the same way, you need to devote ample time when planning for a usability test. At the planning stage, you need to involve a number of key players. Schell (1987) states that—at a minimum—the following people should be involved:

- Writers
- Product designers
- Human factors personnel
- Marketing and sales representatives
- Customer support and quality assurance personnel

The involvement of these people increases the chances that the results of the tests will be accepted and implemented.

Dumas (1989) advocates that a team follow seven steps when planning a test. I describe these steps below, adding suggestions from other sources. You can follow these steps to plan a test, whether you conduct it in a laboratory or in the field. As you read these steps, you can also refer to the extended example that follows. It shows how I used these steps to plan a usability test on subjects' use of the documentation for the AppleShare™ telecommunications network.

Step 1. Identify Test Goals. Discuss as a team your objectives or what each of you wants to learn from the test. (See the earlier section on Determining Test Objectives in this chapter.)

Step 2. Decide Among the Alternative Test Methods. Determine the type of testing you need to perform (from Vogt, 1987):

- *Prototype testing* checks out an early idea or approach. In this sense it is an early use of formative evaluation.
- *Verification testing* makes sure that the documentation can be used by users. Verification testing can be used as part of formative or summative evaluation.
- *Comparative testing* compares two ways of presenting the same information, such as comparing layouts for a manual. Comparative testing is best used as part of formative evaluation.

In determining the type of test, consider whether subjects will use the documentation with the product, use the documentation with a simulator, or use the documentation followed by a pencil and paper test. Also consider whether subjects will use the documentation to learn something (such as the flow of information through a processing unit), to do something (such as complete a form), or to learn to do something (such as how to record a check transaction for access at a later time). You should gather information in a way that best represents how subjects will use the information. For example, a pencil and paper method could be used to get data about subjects' strategies when filling out planning forms, marking up coding sheets, or reading phone bills.

One powerful testing technique, particularly in the case of a single user, is *protocol analysis.* In a protocol analysis, users are asked to vocalize (speak aloud) their thoughts. Protocol analysis is one of the most useful formative tests for studying the usability of a product or documentation because it provides so much information about users' thought processes. Most data from other tests provide little insight into the internal structures of users' thinking as they use products.

You can collect verbal data in two ways:

- *Retrospective verbalization,* in which users talk about what occurred at an earlier time
- *Concurrent verbalization,* in which users talk about their thoughts simultaneously while performing the tasks

Hayes and Flower (1983) describe two types of concurrent verbalization: directed reports, in which users report only specified behaviors (such as talking only when subjects use a help system or only when they don't understand something), and think-aloud protocols in which users utter every thought that comes to their minds.

Using protocol analysis, you collect data both on what users do (such as the sequence of actions taken) and on the processes underlying the sequence of actions taken (such as the goals users have that lead them to complete certain steps, why users want to perform steps in a certain order, and what users are looking for at different stages of using documentation).

Although users can't verbalize all of their thoughts, the gaps between talking also provide important data. Thus, in a protocol analysis, note users' pauses. At these points users may have too much data to process or they may have recognized what task to perform and immediately proceeded to completing the task. Your job is to take the incomplete record from the protocol analysis—together with your knowledge about the nature of the tasks, the product, and the users' abilities to complete the tasks—and infer from these the usability of the information.

Step 3. Identify the Qualifying Characteristics of Test Subjects. Identify typical users for your product and documentation. Rosenbaum (1987) suggests that along with standard user *typologies* (such as job experience, attitudes, typing and reading skills, learning styles, age, and exposure to similar equipment), teams should discuss user profiles for the product (information collected as part of earlier audience analysis) and general user-group statistics (existing statistics on subject populations such as the number of Fortune 500 companies with various types of electronic equipment).

Ramey (1987) suggests that subjects also be grouped according to shared characteristics or experience. To facilitate this process, she has subjects complete a questionnaire at the beginning of the test to confirm their characteristics and experience. This information enables the test team to better interpret the data after the test.

Both Ramey and Rosenbaum state that subjects should not be from the same organization as the developers of the product. Despite their seeming lack of knowledge of a specific product, company employees are too familiar with the company's jargon, terminology, and marketing approach to make them suitable subjects for usability tests. The right subjects may be from companies that will be using your products, from employment agencies, or from colleges and universities.

Step 4. Create Realistic Tasks that Test the Product Design or Manual. Well-designed tasks require subjects to perform the tasks so that performance can be measured against objectives. Vogt (1987) suggests writing test scenarios, each scenario being a self-contained set of actions for a task or set of tasks that subjects are to perform.

For example, if a subject is to test the documentation for a telecommunications network, learning how to use the computer is the main task, and signing on to the telecommunications system is a subtask. In this case, you could write a short scenario for each subtask, and these scenarios could be written explicitly or implicitly.

An *explicit scenario* directs subjects to perform specific actions such as pressing a function key or setting up a product. For example, for a telecommunications network, an explicit scenario might direct subjects to get on the system by going under a specific menu and choosing the menu item that brings up a specific window.

In contrast, an *implicit scenario* creates a set of conditions that results in subjects choosing to perform an action, such as determining what to do to recover from an error. For example, an implicit scenario for a telecommunications network might be to have subjects create a message. Subjects would then likely want to save this message and send it to someone, and they would work to locate instructions to help them do so.

By structuring test scenarios, the test team can monitor how subjects:

- Locate information. Do subjects look up specific information relevant to the scenario?
- Read information. Do subjects read the information required to perform the task?
- Perform the task. Are subjects able to perform the task after reading the information?

Step 5. Order the Tasks. Testing every part of a large manual is difficult. Therefore, you need to order the tasks according to those tasks and information most critical to using the product. Additional testing priorities should be established for tasks that are new or different, occur frequently, or are likely to cause problems.

Step 6. Determine Which Performance Measurements to Take. *Performance measurements* include:

- *Quantitative information,* such as the percentage of subjects who complete the tasks, the amount of time to find information, read information, and do the tasks, the number of errors made and the amount of help given, and the number of times specific pages are used, and
- *Qualitative information,* such as comments from the subjects and comments from the test team.

As a test team, remember to discuss your performance goals for subjects; that is, decide which percentage or number of completed tasks is acceptable. If this is a formative evaluation on an early draft, perhaps 50% will be an acceptable level while users of later versions should reach levels closer to 100%.

Step 7. Create Special Materials Needed to Conduct the Test. Here you can prepare a step-by-step description of everything that should take place as subjects go through each task. For example, this description would include the pages where subjects should find needed information, what subjects need to do to complete each task successfully, and which other sources the subjects should use to get the necessary information. This description helps your test team determine what subjects should be doing at all times during the test (Vogt, 1987).

You should also prepare a test introduction package that explains the subjects' role in the test, how they should handle situations that may come up during the test, and the materials they need to participate in the test (Vogt, 1987). This package could also contain a subject questionnaire (as described in the extended example on pages 319–322).

A final comment on planning the test. Watch out for members of the design team who have been too close to the product and too far away from the users. A sign of a problem occurs when developers cannot articulate the specific characteristics of a potential subject for a test. They may say that the product is intended for "all levels of skill" when the manuals are clearly intended for novice users. Another indication that developers are too close to the product is their propensity to talk about what the product does when you ask them to list the tasks that people will do with the product (Dumas, 1989). To help with these problems, throughout the planning steps, phrase the tasks that users will do in words that relate to the ways users do their work and not in ways that describe how the product works. For example, instead of asking subjects to "Use the 'File Transfer' Option," state that, "Your supervisor has asked that you send this file to Ms. Prestrud in accounting." As Dumas states, "Once developers grasp the concept of task-oriented design, they frequently carry the concept with them to new designs. As a result, you may be able to use future usability tests to fine-tune the usability of a product, rather than to uncover the same serious usability problems over and over" (p. 39).

A Special Note for Planning Field Tests. Conducting usability research in the field involves additional constraints not present in laboratories. Customers can spare only so much time, so you may need to plan for a large number of short visits (Mirel, 1988). The presence of a researcher in the workplace can be disruptive, so you need to explain your research goals to other people in the workplace. You must also observe legal constraints on proprietary information; ethical constraints as you meet, interview, and get to know unique individuals; and budgetary constraints, as it costs money to send technical communicators out into the field (Gould & Doheny-Farina, 1988).

As with laboratory research, you must devote a large percentage of time to planning. According to Gould and Doheny-Farina (1988, p. 335), in addition to structuring the research team and identifying your goals and objectives, you should:

1. Do preliminary research before setting up research situations in the field. Here you should talk with technical support representatives and marketing personnel who already have contacts in the field. You should review the history of user feedback: user surveys, trip reports, complaints, and personal accounts of users' responses to products and information. You should read publications that contain reports on the uses of products and information similar to yours.
2. Put together a research plan. Here the test team needs to talk about the scope of the research: What types of documents, products, and users do you want to study? Which data collection methods will you use? What customer sites will allow investigations? How will you go about setting up formal or informal meetings at the site? What is your company's procedure for getting your research program authorized?

Extended Example: Testing Documentation for the AppleShare™ Telecommunications Network*

1. Test goals
   - We want to learn whether users can locate and use information from two versions of a telecommunications guide to help them complete the following tasks:

™Trademark of Apple Computer Corporation.

* This test was conducted and is reported in Duin, A. H. (1990). This test was conducted in a college classroom and as such was not a true laboratory setting where only one or two subjects might be studied at one time and be videotaped for later in-depth study of users' behaviors.

- □ Get onto the network (the file server).
- □ Check their individual folders/files on the server.
- □ Read a file on the server.
- □ Transfer a file from the server to their individual disks.
- □ Transfer a file from their individual disks to the server.
- □ Get off the server.

- ■ While we would like users to be able to complete each of the six tasks within approximately five minutes, we are more interested in locating trouble spots in the two versions of the documentation.
- ■ We are most interested in deciding which version of the documentation results in greater ease of learning how to use a telecommunications network. Therefore, we will also study users' preferences for the two versions.

2. Test methods

- ■ We will use comparative testing to study the two versions of the documentation (cards versus traditional documentation). Half of the users will receive the cards version of the documentation and half will receive the traditional version. Both versions follow a minimalist approach to designing documentation.
- ■ Subjects will use the documentation while working with the telecommunications network. Afterwards, subjects will respond to a questionnaire on their preferences for the documentation and use of the network.

3. Characteristics of test subjects

- ■ We will test approximately two sections of students currently enrolled in a "computer use" section of a technical writing course (maximum of 80 students). These students represent a cross section of students at the University of Minnesota, and they represent future students who will be asked to use the telecommunications network to aid their collaborative work.
- ■ On the questionnaire that subjects will fill out, we will ask subjects for information concerning their past work with computers and telecommunications.

4. Tasks for testing the documentation

- ■ After a short introduction to the concept of telecommunications, we will give students the documentation and ask them to complete six tasks, which they will need to do throughout the term as they collaborate on technical writing projects. (The six tasks are listed under No. 1 Test Goals).
- ■ Subjects will be given a checklist of the tasks so that they can check off each task once they have completed it. Each task on

the checklist will contain a short, explicit scenario that tells subjects why they need to get onto the network, check their individual folders/files, and so on.

5. Order for the tasks

   - Although the two versions of the manual contain 14 possible tasks, the checklist containing a set of six tasks will function as a way to order the tasks. From interviews with instructors who have taught in the computer lab, we identified these six tasks as being most critical to the use of the telecommunications network.

6. Performance measures

   - Quantitative data will include observations of subjects' computing behaviors. Three trained, independent observers will record the following for each subject:
     - Time spent on each of the six tasks
     - Number of questions asked (simple tally)
     - Number and type of errors made (error categories include mechanical, manual, window, typing, and miscellaneous errors)
     - A rating as to how well the subjects coordinate their use of the documentation with the screen display
     - A rating as to how much subjects enjoy the tasks.
   - Qualitative data will be a questionnaire that will ask subjects about the following:
     - Their attitudes toward the documentation and use of the network
     - Their ease in completing the six tasks
     - Their ease in coordinating their attention between the instructions and the screen
     - Their overall rating of the instructions
     - The percentage of total time they used the help offered by way of the documentation, the on-screen help, other subjects, or their instructor
     - The percentage of their total errors that they feel are attributed to mechanical, instructional, window, typing, or miscellaneous errors
     - How the documentation compares to other computer instructions they have used in the past
     - How much they intend to use the network to communicate with their instructor and group members and/or to store their own work in their private folders on the system

- Any other comments about their use of the network and its accompanying documentation.

7. Test Materials

- The three observers will meet before the test to go over the recording materials and to conduct a pilot test while using the materials. We will then talk about differences in their interpretation of the scales for recording subjects' computing behaviors. These observers will not be part of the team that has designed the documentation. In this way, they will be independent observers who will more likely discover valid usability concerns.
- All subjects for the test will receive a *Human Subjects Form* that they can read before the usability test. This form will indicate exactly what they will be asked to do during the test, and if a subject is willing to be a test subject, he or she will sign the form.

How to Conduct a Usability Test

The procedures for conducting a usability test vary depending on whether you conduct it in the laboratory or in the field.

Laboratory Tests. Tests conducted in a laboratory obviously require a test lab or a test site. Figure 11.1 presents a test laboratory at the American Institutes for Research. This lab contains the following components:

- A reception area where subjects can become acquainted with the test team.
- An observation room where the test team can log data while observing the test through the one-way mirror.
- A test room which is as realistic as possible. For example, if the subjects are receptionists, perhaps you could forward real office calls to the site so that typical phone interruptions take place during the test.

Before the actual test, you should run a *pilot test* with subjects so that you can evaluate:

- The tasks selected, to make sure they are appropriate and can be tested.
- Scenarios, to make sure they are realistic and users can complete them.
- Materials, to make sure they can be used.
- Laboratory equipment, to make sure it works properly.
- The test team's readiness to conduct the test.

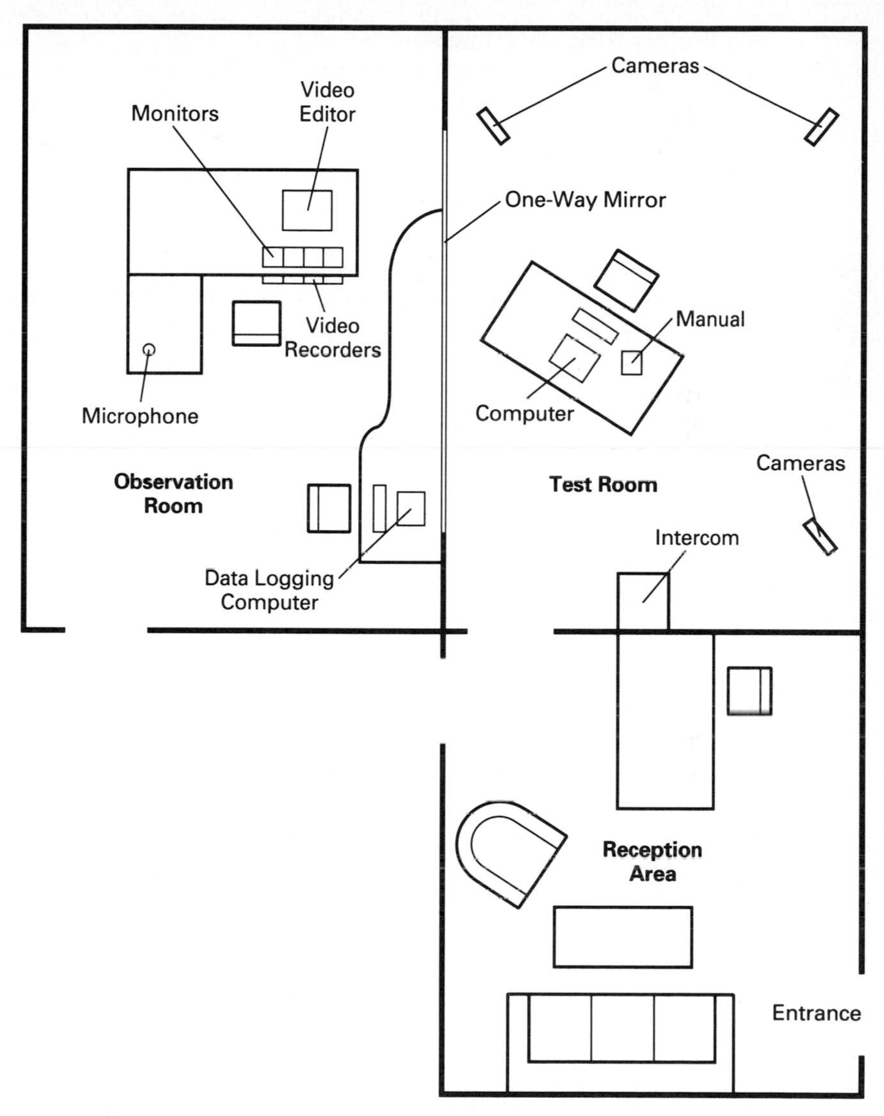

FIGURE 11.1 A Usability Test Laboratory at the American Institutes for Research (© Copyright 1989, American Institutes for Research)

By running a pilot test, you learn how subjects interpret the scenario you have written and whether you have selected too many tasks to complete during the test time. Pilot testing also allows you to adjust recording techniques, to check the product and testing equipment to make sure both are working correctly, to check that all recording codes are understood by the test team, and to practice how subjects will be greeted and introduced to the lab site and the usability test.

During the actual testing, you'll have the best control if you have followed the planning steps discussed earlier and can do the following:

- Confirm subjects' characteristics.
- Control the presentation of information through written materials.
- Ensure that each subject's data will be recorded consistently, that troubleshooting processes are in place, and that outside influences—such as the number of breaks subjects take—are controlled through your step-by-step description or checklist of the process.

Field Testing. When conducting field tests, you need to discover if, when, and how people use products and information on the job. To learn this, you use a variety of methods; this variety of methods is often called *methodological triangulation* (Doheny-Farina & Odell, 1985). Gould and Doheny-Farina (1988) suggest three main forms of collecting data: questionnaires, interviews, and on-site observations. Mirel (1987) also suggests a three-pronged method of inquiry consisting of user logs, longitudinal observations, and interviews and/or surveys. In the following discussion, I note how four organizations are using a variety of these methods for collecting quantitative and qualitative data on the usability of products and documentation in the workplace.

As mentioned before, engineers at Digital Equipment Corporation (DEC) observe users actually working with DEC computer systems, and they conduct interviews after users have been observed. They interview a variety of users and look for common elements of usability for groups of people as well as distinctive elements of usability for individual users. The interview process continues until new interviews no longer reveal much new usability data or until resource and time limitations stop the process. DEC's approach is to start with a small number of interviews (four or less) with users in various jobs. These interviews then become prototypes to determine how many and what types of users will be most helpful for uncovering new data. The engineers videotape or audiotape each session, and the team members meet after each interview to reconstruct an accurate record of the events (Good, 1989).

At the University of York, Wright and Monk (1989) advocate a quantitative and qualitative approach that involves *system logs* and verbal protocols. They first identify problems from system logs collected at the field site. Second, they analyze data logs from users who complete tasks set up by the field researchers. Third, they use *reenactment* in which testers recreate the problem by using the system logs, and users and testers discuss the problem as they view the system logs. Last, researchers question users to learn what they know about how the system works or what the system can do.

Human factors consultants at Pacific Bell conduct field research to identify difficulties, confusion, and dissatisfaction that users experienced while interacting with computer systems. They also record users as they use a system in their workplace, asking users to think aloud as they interact with the system (protocol analysis).

Anderson (1989) states that during field testing, the testers work hard to remain observers and not interviewers. Despite their planning, Anderson notes that field tests present problems: they lack quantification and control over the tasks that users perform and the environment in which users work, and they involve a limited number of subjects. For example, a subject in a test laboratory naturally will not have the same interruptions that would likely occur in the subject's workplace, and sophisticated recording systems are usually difficult to use in the subject's workplace. As a result, some organizations resist field tests.

Responding to this same resistance, field researchers from IBM state that:

> Quantitative measures, like keystroke counts and performance times on benchmark tasks, leave out most of what is important about usability: the kinds of errors that users make, their satisfaction or dissatisfaction with features of the interface, etc. Such information can only be obtained through qualitative measures, such as various forms of interviews, thinking out loud, and video observation. When used in the laboratory, qualitative methods generate much useful information about users' interactions with systems, but in an artificial environment in which users work on contrived tasks for limited amounts of time at an unrealistic pace (Campbell, Mack, & Roemer, 1989, p. 30).

One area in which IBM is using field methods is with *task-oriented* online assistance. Using keyboard-activated tape recorders, users make comments while using the help for a text editor. IBM researchers also conduct interviews with several sources to develop a composite picture of the users' work. Then, researchers summarize their interpretations of the interviews in a narrative. This narrative focuses on procedural aspects of work (such as the tools used and characteristics of the work process), on what users need to know or have or do to accomplish goals, and on the social structure of the users' work environment. This task-oriented narrative provides support for changes in the design of the product and documentation.

In each of the above cases, researchers used interviews as a crucial method in their usability research. Gould and Doheny-Farina (1988, p. 338) further expand on three types of interviews that researchers can conduct in field settings:

- *Post-hoc interviews* (Rose, 1984) let you closely observe users working with a product for the first time. You take detailed notes, identifying points when users pause, stumble, make errors, or get stuck. Immediately after the session, you interview users by reviewing your notes and asking users to recall their thoughts about the pauses or errors.
- *Discourse-based interviews* (Odell, Goswami, & Herrington, 1983) let you identify documentation that has been rewritten by users and use these rewrites as the basis for the interviews.

- *Scenario-based interviews* let you propose various situations in which users would operate the system, fix it, or recover from errors and ask users how they would address the situations. In these interviews, you discover preconceptions and strategies that users bring to the tasks.

In all of these methods, you need to record information either through field notes or recording devices. After collecting the data you should:

- Review all of the responses.
- Transcribe all verbal information.
- Make an index of your notes so that other teams can understand your findings and build on them.

Glaser and Strauss (1967) suggest that you do the following:

- Read the data chronologically, searching for patterns.
- Establish categories that link the patterns together.
- Develop themes that link the categories and explain the users' behavior.

HOW DO YOU REPORT THE DATA COLLECTED IN A USABILITY TEST?

Usability research from both laboratory and field settings yields a substantial amount of data. One laboratory test may produce a 15- to 20-page data log that includes each action of the subject, the videotapes of the subject, notes made by members of the test team, notes summarizing the post-test interview, background information on the subject, and post-test questionnaires (Dumas, 1989). Likewise, field testing on one user in the workplace may include keystroke logs, videotapes or audiotapes, interview and survey data, and notes on the social environment of the workplace. In both cases, the analysis and packaging of this amount of data is the key to stimulating short- and long-range changes in the design and development of products and documentation. Dumas (1989) states that "The way you report this data will have a major influence on the impact of the test. *If you focus only on the specific problems in the product you test, the product design will be improved, but you cannot be sure that future designs will be better*" (p. 40).

Both Dumas (1989) and Redish (1989) suggest setting a *severity code* for each problem that has been discovered. A severity code indicates the priority the problem should receive:

- Level 1—a problem that prevents the user from performing or completing a task.

- Level 2—a problem that causes significant frustration for the user.
- Level 3—a problem that does not cause significant frustration, but a change would make the task easier.
- Level 4—a problem that can be fixed in future versions.

In reporting these problems, Dumas states that you should make it clear that these problems may not be the only problems nor the most serious ones.

You report problems in a formal test report. According to Schell (1987), the test report should contain the following:

- An introduction that presents an overview of the testing activity and a summary of the test results.
- A complete description of the test.
- Objective test data (such as quantitative data from keystroke results or trained observers).
- Subjective test data (such as qualitative data from questionnaires and interviews).
- Conclusions and recommendations.
- An appendix including all written test materials (such as the actual questionnaires, test scenarios, forms, or checklists).

Field testing can also be reported following this framework. However, this type of testing is often reported in the form of a trip report that describes the following (Gould & Doheny-Farina, 1988, pp. 341–342):

- User problems (documentation problems and product design problems).
- Examples of re-worked or annotated documentation.
- Successful documentation, design strategies, or features.
- The nature of the workplace (including maps of user and documentation locations).
- The strategic role that information plays within the company in terms of the user tasks.
- Physical use of documents (how they are accessed and used).
- Users' views of their jobs.
- Your testing methods.
- Reports of problems that may have influenced the validity of the report (attitudes toward your company, limited access to certain user groups).
- Need for further research.

In any such report, consider it as more than a record of results. Think of it as an agent for change that can lead to changes in the particular product and its documentation, to new standards and guidelines for future work, and to revised roles and procedures within the company.

REVISITING THE COMMUNICATION PROBLEMS

Consider again the problems discussed at the beginning of this chapter. The Candle Corporation, which needed to revise a system for new audiences, chose to revise the menus. Because the menus were not designed according to the job tasks of these audiences, the menus were not being used (Tavlin, 1989). The test team included only managers from Research and Development, which most likely did not represent a broad enough test team. Developers and technical communicators need to be added to this team. In this case, too, the product was already out in the workplace and was used by diverse audiences. It would seem that field tests would help Candle Corporation learn more about its users, the environments they work in, and most of all, for what tasks these people used the system. While this field testing would act first as an evaluation of the current product, it could also provide ideas for future products. Usability testing in a laboratory setting would also help in this situation. The company could assess subsets of the 800 commands. Laboratory testing with new customers could uncover those tasks and commands most pertinent to new users, and these commands might well be a different set than those used by the previous users.

The manual developed at the Illinois Institute of Technology was rarely used by cashiers even though it supposedly went through numerous laboratory tests (Mirel, 1987). The laboratory tests evaluated the performance of the system or the users' interaction with the system, but not the manual. Although laboratory tests centered on users' tasks might have shown a very usable product, the laboratory setting did not uncover three factors that Mirel discovered from her field testing on the use of the manual: that users were inhibited from using the manual because of

- Time constraint.
- Reliance on interpersonal exchanges to maintain social cohesion in the office.
- A supervisory system in which the bursar acted as the authoritative reference point for all complicated problems.

This type of data can only be collected in the field.

Last, the Department of Defense still fails to provide usable manuals even though writers follow an immense structure of guidelines and specifications based on document design research. From their research on the production process of these technical manuals, Duffy, Post, and Smith (1987) found that validation or testing of the documents is the problem. Testing consists only of the contractor's certification that the manual has been validated and is accurate. This validation procedure could range from "tabletop validation" in which someone reads through the manual to an informal process in which peer engineers and technicians already familiar with the system would perform tasks while the writers observed. In essence, the validation or testing process rarely includes evaluation by actual users. This case clearly shows that documentation that follows all the latest guidelines and rules will ultimately fail if it is not tested with *real* users.

Could testers go "out into the field" of military personnel and observe the actual use of the manuals? Yes.

Could army publication houses set up laboratory test sites? Yes. This initially costs money, but in the end it can save much more than it costs.

Researchers have long known that review and revision by a writing expert is an inadequate means for assessing the usability of information in a document (Swaney, Janik, Bond, & Hayes, 1981). Improvement comes only when the review and revision is based on actual testing with users.

To conclude, techniques for evaluating the usability of documents are not failsafe because each is sensitive to a particular dimension of users' problems. Since all usability testing techniques risk failing to detect users' problems lying in other dimensions, it is much safer to use a variety of procedures rather than relying on a single technique (see Wright, in press). Therefore, do not choose simply to do only laboratory testing or field research. Develop a combination of testing techniques and work in a variety of testing environments. Your choice of techniques and environments should relate to the needs of your users and your test team, and these techniques should help you develop a more systematic form of usability testing that will drive your future information designs and decisions.

SUMMARY

Usability testing is a procedure in which people who are representative of the intended users of a product are asked to use a product and its information. In testing, technical communicators study the users' degree of effectiveness, efficiency, and comfort when using the product and its information. You should conduct usability tests to make information usable, improve products, remove obstacles, and save costs.

You can conduct a usability test in a laboratory or in the field. In a laboratory, you construct an environment like that of intended users and watch them use the product and its information. Because the environment is controlled, you can see the causes of users' problems as well as how they respond to the problems. In the field, you watch people use a prototype or the actual product and its accompanying information in the workplace.

You conduct a usability test as follows:

1. Plan the test.
   a. Determine your objectives or what you want to learn from the test.
   b. Decide on the type of testing you need to perform.
   c. Identify typical users for your product and information.
   d. Create realistic tasks that test the product design and accompanying information.
   e. Order the tasks according to what tasks are most critical to using the product.
   f. Determine quantitative and qualitative measurements and discuss what equals an acceptable level for completed tasks.
   g. Prepare all special materials needed to conduct the test.
2. Conduct the test.
   - If in a laboratory, run a pilot test with subjects to evaluate your tasks, materials, equipment, and your test team's readiness to conduct the test.
   - If in the field, use a variety of methods such as questionnaires, interviews, and on-site observations. In field testing, you should study how people use the products and documentation, you should identify the sources of information that users prefer for obtaining help, you should isolate the problems that provoke users to seek information, and you should analyze users' perceptions of the product and documentation.
   - In both laboratory and field testing, you need to record information either through field notes or recording devices.
3. Review and report the results.
   - In laboratory testing, review the data and attach a severity code for each problem discovered. In reporting the results, include a complete description of the test, all quantitative and qualitative test data, and an appendix that includes all written test materials.
   - In field testing, transcribe all verbal information, make an index of your notes, and then begin searching for patterns in the data and establishing categories that link the patterns and that explain the users' behaviors. In reporting the results, follow a framework such as a trip report that includes information such as users' problems, examples of reworked or annotated documentation, the nature of the work environment, your methods, and problems that may have influenced the validity of your research.

LIST OF TERMS

comparative testing Compares two ways of presenting the same information, such as two layouts for a manual. Best used as part of formative evaluation.

concurrent verbalization Method of verbal collection in which users talk about their thoughts simultaneously while performing tasks.

explicit scenario Directs test subjects to perform specific actions such as pressing a function key or setting up a product.

field tests Usability tests or evaluations conducted in the workplace of the subjects who use the product and accompanying information.

formative evaluation Helps designers and writers both evaluate the product or documentation and apply the results as they redesign and refine it.

human subjects form Type of consent form that subjects sign prior to participation in a study. The form indicates what subjects will be asked to do, whether the test could result in any harm to them, and whether the resulting data will identify them in any way.

implicit scenario Creates a set of conditions that results in subjects choosing to perform an action, such as determining what to do to recover from an error.

laboratory tests Usability tests conducted in a constructed environment similar to that of the intended users of a product and its accompanying information.

longitudinal observation Observing users over an extended period of time.

methodological triangulation Using a variety of methods to study users; not relying on only one method or testing technique.

performance measurements Quantitative and qualitative data.

pilot test Test conducted before the usability test to check the product and testing equipment to make sure both are working correctly, check that all recording codes are understood by the test team, and practice how subjects will be greeted and introduced to the lab site and the usability test.

post-hoc interviews After observing users working with a product for the first time, you interview users by reviewing your notes and asking users to recall their thoughts when using the product and documentation.

protocol analysis Users vocalize (speak aloud) their thoughts. Most useful in formative testing for providing insight into the internal structures of the users' thinking about the product or documentation.

prototype testing Early form of formative evaluation that tests an idea or approach early in the development cycle.

qualitative information Data that cannot be measured according to a specific scale.

quantitative information Data that can be measured according to a specific scale.

reenactment Verbal protocol method to learn about users' plans and goals. In reenactment, tests of the problem are recreated by using the system logs, and users and testers discuss the problem as they view the systems logs.

retrospective verbalization Method of verbal collection in which users talk about what occurred at an earlier time.

scenario-based interviews After proposing various situations in which users would operate a system, fix it, or recover from errors, you interview users, noting the preconceptions and strategies that they would bring to the tasks.

severity code Indicates the priority that a problem discovered during a test should receive.

summative evaluation Assessing the impact, usability, and effectiveness of the product and accompanying documentation. Summative evaluation mainly includes quantitative measures and checks for verification. The product and its documentation are tested at the end against preset criteria, and they either pass or fail.

system logs Data collected from computer programs that record users' interaction with the system.

task-oriented Information that accompanies a product and is structured according to the tasks that users need to perform rather than according to the functions performed by the product.

test team People involved in planning, implementing, and reporting the usability test.

typologies Ways to characterize users such as job experience, age, or education.

usability Degree to which an intended audience can perform the desired tasks in the way those tasks are usually performed.

usability test Procedure in which people who are representative of the intended users of a product are observed using the product and its information.

verification testing Makes sure that the documentation can be utilized by the product users. Can be part of formative or summative evaluation.

REFERENCES

Anderson, R. (1989). Notes about some experiences with contextual research. *SIGCHI Bulletin, 20*(4), 29–30.

Booth, P. (1989). *An introduction to human-computer interaction.* Hillsdale, NJ: Erlbaum.

Brooks, T. (1991). Career development: Filling the usability gap. *Technical Communication, 38*(2), 180–184.

Campbell, R. L., Mack, R. L., & Roemer, J. M. (1989). Extending the scope of field research in HCI. *SIGCHI Bulletin, 20*(4), 30–32.

Doheny-Farina, S., & Odell, L. (1985). Ethnographic research on writing: Assumptions and methodology. In L. Odell & D. Goswami, (Eds.), *Writing in non-academic settings* (pp. 503–535). New York: The Guilford Press.

Duffy, T. M., Post, T., & Smith, G. (1987). Technical manual production. *Written Communication, 4*(4), 370–393.

Duin, A. H. (1990). Computer documentation: Centering on the learner. *Journal Of Computer-Based Instruction,* 17, 73–78.

Dumas, J. S. (1989). Stimulating change through usability testing. *SIGCHI Bulletin, 21*(1), 37–44.

Glaser, B., & Strauss, A. (1967). *The discovery of grounded theory: Strategies for qualitative research.* New York: Aldine.

Good, M. (1989). Contextual field research in a usability engineering process. *SIGCHI Bulletin, 20*(4), 25–26.

Gould, E., & Doheny-Farina, S. (1988). Studying usability in the field: Qualitative research techniques for technical communicators. In S. Doheny-Farina (Ed.), *Effective documentation: What we have learned from research* (pp. 329–343). Cambridge, MA: The MIT Press.

Grice, R. A. (1991). Introduction: Making information usable. *Technical Communication, 38*(2), 178–179.

Hayes, J., & Flower, L. (1983). Uncovering cognitive processes in writing: An introduction to protocol analysis. In P. Mosenthal, L. Tamor, & S. Walmsley (Eds.), *Research in writing: Principles and methods* (pp. 207–220). New York: Longman.

Mirel, B. (1987). Designing field research in technical communication: Usability testing for in-house user documentation. *Journal of Technical Writing and Communication, 17*(4), 347–354.

Mirel, B. (1988). The politics of usability: The organizational functions of an in-house manual. In S. Doheny-Farina (Ed.), *Effective documentation: What we have learned from research* (pp. 277–297). Cambridge, MA: The MIT Press.

Odell, L., Goswami, D., & Herrington, A. (1983). The discourse-based interview. *Research on writing.* New York: Longman.

Queipo, L. (1991). Taking the mysticism out of usability test objectives. *Technical Communication, 38*(2), 185–189.

Ramey, J. (1987). Usability testing: Conducting the test procedure itself. *Proceedings of the 1987 IEEE International Professional Communication Conference* (pp. 127–130). Winnepeg, Manitoba.

Redish, J. C. (1989, September). *It works. Fine. But can people use it?* Keynote presentation at Technicom, Toronto.

Redish, J. C., & Schell, D. A. (1989). Writing and testing instructions for usability. In B. E. Fearing & W. Keats Sparrow (Eds.), *Technical writing: Theory and practice* (pp. 63–71). New York: The Modern Language Association of America.

Rose, M. (1984). *Writer's block: The cognitive dimension.* Carbondale, IL: Southern Illinois University Press.

Rosenbaum, S. (1987). Selecting the appropriate subjects: Subject selection for documentation usability testing. *Proceedings of the 1987 IEEE International Professional Communication Conference,* (pp. 135–142).

Schell, D. A. (1987). Overview of a typical usability test. *Proceedings of the 1987 IEEE International Professional Communication Conference* (pp. 117–120). Winnepeg, Manitoba.

Selzer, J. (1983). The composing processes of an engineer. *College Composition and Communication, 34,* 178–187.

Shriver, E., & Hart, F. (1975). *Study and proposal for improvement of military technical information transfer methods.* Aberdeen Proving Grounds, MD: U. S. Army Human Engineering Laboratory.

Sullivan, P. (1988). Users and usability: On learning about users. *Proceedings of Sig Doc* [October, 1988]. Ann Arbor, MI.

Swaney, J. H., Janik, C., Bond, S., & Hayes, J. R. (1981). *Editing for comprehension: Improving the process through reading protocols* (DDP TR-14). Pittsburgh, PA: Carnegie-Mellon University.

Tavlin, B. (1989). Experience with the design of a real-time computer performance monitor. *SIGCHI Bulletin, 20*(4), 27–28.

Vogt, H. E. (1987). Designing the test—Writing the scenarios. *Proceedings of the 1987 IEEE International Professional Communication Conference* (pp. 121–126). Winnepeg, Manitoba.

Winograd, T., & Flores, F. (1986). *Understanding computers and cognition: A new foundation for design.* Norwood, NJ: ABLEX.

Wright, P. (in press). Is evaluation a myth? *Information design papers.* London: Architecture Design and Technology Press.

Wright, P. C., & Monk, A. F. (1989). The elicitation and interpretation of experiential data within an iterative design framework. *SIGCHI Bulletin, 20*(4), 26–27.

SUGGESTED READING

Claparede, E. (1934). Genese de l'hypotheses (Genesis of the hypotheses). *Archives de Psychologie, 24,* 1–55.

Duncker, K. (1926). A qualitative (experimental and theoretical) study of productive thinking (solving of comprehensible problems). *Pedagogical Seminar, 33,* 642–708.

Eason, K. D. (1984). Towards the experimental study of usability. *Behavior and Information Technology, 2*(4), 357–364.

Ericsson, K. A., & Simon, H. A. (1979). *Thinking-aloud protocols as data: Effects of verbalization.* C.I.P. Working Paper No. 397.

Schumacher, G. M., & Waller, R. (1985). Testing design alternatives: A comparison of procedures. In T. M. Duffy, & R. Waller (Eds.), *Designing usable texts.* Orlando, FL: Academic Press.

Smagorinsky, P. (1989). The reliability and validity of protocol analysis. *Written Communication, 6*(4), 463–479.

CHAPTER 12

PROTECTING YOUR WORK: PROFESSIONAL ETHICS AND THE COPYRIGHT LAW

Jacquelyn L. Monday

Publications Chief, Louisiana Geological Survey

Mary C. Hester

Technical Writer and Editor

> Jane is a technical editor working for an environmental consulting firm. The company conducts research for private parties, scientific foundations, and government agencies on the potential impacts of proposed housing and commercial developments, highways, dams, and other activities. In assembling the proposals to conduct the research and the reports on completed research, Jane has occasion to see some facts, figures, and illustrations used repeatedly. She notices that sometimes these uses include citations to the sources of the information, and sometimes they do not. When she asks one of the scientists to provide a source for an item that she is fairly certain has already appeared somewhere else, he tells her, "Oh, these diagrams and maps have been used so many times, they don't really belong to anyone anymore. No one pays attention to those sorts of details anyway." This strikes Jane as wrong: what is the use of doing scientific or technical work in the first place if people can later just use it any way they want, without giving you credit for it? Don't people who do this sort of technical work have any right to it?

The answer to Jane's musings is that there is, in fact, a set of rights, obligations, and guidelines that governs the use of original works of authorship created by people in the United States and most other countries. Although in practice the various rules and standards are interdependent, they can be thought of as falling into two distinct systems: the laws of copyright (which primarily protect *ownership*) and the rules of professional ethics (which primarily protect *originality*).

The legal protection and ethical treatment of the written, visual, and computerized information that technical communicators work with every day are two of the most serious and complex issues we face. This chapter specifically addresses the following questions and situations that arise from those issues:

- What is the nature of the rights and obligations, both legal and ethical, technical communicators face in creating their own documents and in dealing with others' works?
- Who is protected by these rights?
- How can technical communicators avoid infringing on the rights of other authors?
- How can technical communicators protect their own original work, both before and after it is published?

YOUR RIGHTS AND OBLIGATIONS IN USING WRITTEN MATERIAL

The legal rules governing the rights and obligations you face in creating documents and in dealing with others' work are set forth in the copyright laws. The protection that copyright law provides lies in the rights it specifically grants with regard to original work. These include the exclusive right of the copyright owner to (1) reproduce the work in copies, (2) distribute copies of the work, (3) prepare derivative works based upon the copyrighted work, and (4) display or perform the work. All types of information—printed, video, or computerized—are covered by the legal provisions. These legal rights are similar to property rights; thus, a copyright can be thought of as a piece of property—hence the concept of ownership.

The foundation for the United States copyright law was laid in Article I, Section 8, Paragraph 8, of the United States Constitution, which gives Congress the power to reserve to authors and inventors the rights to their writing and discoveries. Since then the law has been enlarged and clarified through both statutes and case law. The principles and specifics of the laws that evolved were made explicit in 1976, when Congress passed the Copyright Act of 1976, 17 U.S.C. sec. 101 et seq. With the enactment of the Berne Convention Implementation Act in October 1988, the United States joined the nations adhering to the Berne Convention for the Protection of Literary and Artistic Works, the primary international copyright convention. This action gave U.S. authors enhanced copyright protection in more nations than previously; it greatly increased the value of our country's intellectual property on international markets (Latman, Gorman, & Ginsberg, 1989); and it resulted in modifications to the law as set out in the 1976 act, some of which are discussed in this chapter.

In creating documents or artistic works and in dealing with those created by other people, you face another set of rights and obligations—ethical standards. These are strictures considered binding by all true professionals—whether they be scholars, scientists, writers, editors, artists, or students. The protection afforded by these standards lies in the kinds of fraudulent behaviors they discourage. Chief among these is *plagiarism,* but also included are falsifying information or data, simultaneous submission of the same manuscript to different journals (when the journal opposes this practice), fragmentary or divided publication of the same information in different places, abuse of coauthorship, and repetitive publication of the same information without giving proper credit or obtaining required permissions (*self-plagiarism*).

All of these behaviors are unethical because they undermine the assumption of originality. Plagiarism is an obvious example—it is generally well instilled in high school and college students that it is wrong to present someone else's words, ideas, or data as though they were your own. Likewise, fabricating information and presenting it as fact is clearly wrong. Most of the following discussion applies most directly to the creation and publication of nonfiction academic work, particularly scientific or technical research, but the concepts of honesty and originality are equally pertinent in other fields.

Simultaneous submission of the same manuscript to different publishing outlets in the academic and scientific worlds is generally considered unethical because of the assumption that material appearing in scholarly periodicals is being presented publicly for the first time. Most such journals explain in their information for authors that "submission" means that the material has not yet been published and that it is not being considered for publication by any other journal. In the commercial realm, materials may be submitted to multiple prospective publishers, but only with express notification to each of them that other outlets are being sought simultaneously. Fragmentary publication, in which "research is published piecemeal so as to lengthen a list of personal citations" (Smith, 1985, p. 1292) is fraudulent for a similar reason: readers and colleagues assume that, in general, a piece of original research or other work will be "released" in a single, coherent description of the findings. Releasing the information or results of a project in several separate publications rather than as a unified whole gives a false impression that numerous, different projects were done (of course, that is exactly the impression the unethical person wants to give). On the other hand, some projects, especially multidisciplinary ones, comprise several distinct components that lend themselves to separate treatment in the published literature. In any case, a false impression can be avoided simply by explaining the context and extent of the original work or idea and citing previously published work based on it.

Repetitive publication of the same information without citation is unethical for two reasons. First, it is usually a violation of copyright, which is generally held by the first publisher. Second, it implies that the information was generated as original material for that particular appearance in a publication, when, in fact, it merely repeats or recasts previously published material. An author who does this is plagiarizing his or her own work. To act ethically in reprinting material that you have previously published, you must obtain the publisher's permission, provide a complete citation to the work in which the material was originally published, and, if you have coauthors, obtain their permission.

Abuse of coauthorship generally takes the form of including as coauthors persons who have not made a direct substantial contribution to the work. These persons have been called *honorary authors* (Stewart & Feder, 1987, p. 210), and they may have been added to the list of authors to compensate them for providing the real authors with funding or other administrative support. If the honorary authors are well-known, their names may be added to make the work appear more creditable or to facilitate publication. Because honorary authors are, by definition, not directly involved in acquiring and interpreting data for the project and have not made any essential intellectual contribution to it (Stewart & Feder, 1987), their inclusion promotes "an unwarranted assumption of credit," and is "tantamount to plagiarism by the honorary authors" (Zen, 1988, p. 292).

Professional Codes of Ethics

To establish clear standards of behavior and reduce unethical conduct, many professional organizations have established codes of ethics by which members agree to be bound. Members of the Society for Technical Communication, for example, commit themselves, among other things, to "excellence in performance and the highest standards of ethical behavior," including precise use of language and visuals, taking responsibility for the audience's understanding of the message communicated, and respecting the work of colleagues (STC, 1982). In some countries the professional codes have been included in the copyright laws. These special provisions, which provide for the "moral rights" of authors, are intended to protect authors from certain acts that could injure their personal identities or reputations. They give authors the legally enforceable right to have their names appear on copies of their work, to prevent the attribution to them of other people's work, and to prevent the reproduction of their work in distorted or degrading forms (Latman et al., 1989). In some countries these rights survive an author's death, and the author's heirs or representatives may enforce them (Latman et al., 1989). Although the United States has never included moral rights as part of copyright protection, general provisions of *common law* (that is, law based upon precedents established in courts rather than on legislative action) "such as those relating to implied contracts, unfair competition, misrepresentation, and defamation" offer similar legal protection (Latman et al., 1989, p. 13).

Penalties

The rules of professional ethics and the laws of copyright not only set up standards, but they also provide penalties for people who violate them and relief to people injured by such violations. The penalties that may be imposed in copyright cases are specified in the statute. They range from damages of $200 for innocent (unintentional) infringement to $200,000 for willful infringement. The defendant may also have to pay the costs of actual damages and loss of profits, if these are proven during the lawsuit. The penalties don't end there. Courts can require the defendant to relinquish or destroy copies produced illegally, destroy the printing plates, refrain from future actions that may be construed as infringement, and pay the plaintiff's legal fees.

Consequences in the professional realm likewise vary in form and severity. Private corporations, universities, research institutes, and local, state, and federal government agencies all have procedures for internal investigations of suspected wrongdoing, whether it be plagiarism, copyright infringement, or other forms of fraudulent writing or publishing. The penalties levied by these bodies can range from a mild reprimand to loss of a job—even a tenured position. Even if the suspicions are not proven, the damage to the reputation of the accused (particularly if the investigation is made public) can be so serious that the person may need to seek a position elsewhere or even change careers. It is easy to see that an author of work proved fraudulent is subject to severe repercussions, "usually, total derailment of a career" (Koshland, 1987, p. 141).

Editors of professional journals and publishers, too, can impose significant sanctions: refusing to publish material suspected of being unethically or illegally presented or blacklisting an author who has been known to act unethically. In 1982, the editor of *Science* cited specific precedents for penalties. The editors of *Cell* and of *Proceedings of the National Academy of Sciences* had agreed that for three years they would not publish any manuscripts submitted by a particular author who had previously published similar articles in those journals (Abelson, 1982). An even harsher penalty was invoked against a researcher who had submitted the same manuscript to two journals—Abelson (1982) reported that all U.S. journals in the researcher's field "agreed never to consider any manuscript from that investigator's laboratory" (p. 6251).

WHO IS PROTECTED BY COPYRIGHT LAWS AND PROFESSIONAL ETHICS?

Who is the "author"? Who owns the copyrights? Exactly who is protected by professional ethics? This section explores these questions, explaining when authors own copyrights and when employers do, who owns copyrights when several authors contribute to a work, and when copyrights can be transferred to others.

Whether the Author or the Employer Owns the Copyright

Although both ethical rights and statutory copyright ownership may seem clear-cut, special rules apply to several common situations. In the case of a person who creates an original work as part of his or her responsibilities as an employee, the creation is considered a *work made for hire,* and the copyright is owned by the employer, not the employee—even though the employee created the work. Work made for hire refers both to work prepared by employees within the scope of their employment and to work specially ordered or commissioned, if specific requirements are met. For example, when a scientist employed by a research institute writes a report for a supervisor, the institute is legally the author of the report (Strong, 1982). The conditions required for work done by freelancers to be considered work for hire include requirements that the freelancer and the client sign a written contract expressly stating that the work is being done as a work for hire and that the work fall into one of specifically listed categories. Some of the categories specified are contributions to collective works (such as magazines or anthologies), supplementary work (such as diagrams illustrating work by another author), and atlases.

When Several Authors Contribute to a Work

There are additional considerations when a work has more than one author. Under copyright law, the authors of a joint work are co-owners of the copyright; each holds an "undivided interest in the whole" (574 F2d 476 at 477 [1978]). On the other hand, if each author was wholly responsible for creating a specific portion of the work, as, for example, a collection of separate papers or articles, the authors may agree that their copyrights are divided according to who created which portion. In that situation, the *copyright notice* must name the separate copyright holders.

Although under the law all the authors of a joint work have identical ownership rights, in noncommercial matters the senior author (the author whose name appears first) is often deferred to by the other authors in decisions about how the work will be used, about matters governing its originality, or about professional or scientific responsibility for the information in the document. Although other authors may defer to the senior author, editors assume that any of the joint authors can speak authoritatively about the content of the work and for the other authors.

The question of the relative contributions of multiple coauthors, especially in scientific writing, has given rise to a debate about what authorship really means, and what being listed as an author should suggest to prospective publishers, and, above all, to readers. There is no question that each coauthor need not have participated equally in the work being reported or in the actual writing, but there is no clearcut definition of how much "involvement" is enough to ethically qualify someone for author status. It has been suggested that an *author* is a "person who can take public responsibility for the content" of the paper, report, or other work (Huth, 1982). This means that each *coauthor* should be able and willing to defend, in a forum such as letters to the editor or written debate in professional periodicals, the data, analysis, conceptual framework, methods, interpretation, and conclusions of the work. This does not mean that, for instance, in the case of a multidisciplinary work, one professional should have a thorough mastery of all the details of the others' specialties, but all should have enough understanding of each component and the overall design that they should have to refer only very specific questions to the others.

When Copyrights Can Be Transferred

Like other property, copyrights may be sold or otherwise transferred. In scientific and scholarly settings, this typically occurs when authors release their copyrights to publishing companies or journals that will be publishing the work. Such releases can take various forms, from requiring transfer of all present and future rights to transfer of limited rights, sufficient to protect the journal or publisher for a single edition or issue.

Each author of a joint work has an independent right to use, license, or bequeath the whole work, subject only to a duty of accounting to the other co-owners for the proceeds (Nimmer & Nimmer, 1989, sec. 201). Such a license would not be exclusive, however, because any of the other owners could also grant licenses for the same work. So, for a collection that has separate authors for each section, who also hold separate copyrights for their contributions (a compilation of stories, papers, or essays, for example), one author may not single-handedly transfer everyone's copyrights to another party.

As authors, technical communicators may find themselves in the position of transferring their own copyrights to other people or to a journal or publisher. As editors, technical communicators more often are in the position of asking (or requiring) others to relinquish their rights to a piece of work in exchange for its being published.

HOW TO AVOID INFRINGING ON THE RIGHTS OF OTHERS

> Jane goes to the local library and investigates her suspicions about the propriety and legality of using other people's work without giving them credit for it. After several hours of examining handbooks on copyright law, reading descriptions of lawsuits, and finding out about all the trouble someone could get into by ignoring ethical and legal standards, she throws up her hands in despair. It seems impossible to avoid infringing upon other people's rights to their work.

Jane's reaction is understandable, but it is not really warranted. Fortunately, over the years ethical and legal conventions have developed to allow people to make reasonable use of each other's work without incurring legal or professional sanctions. If Jane follows these rules (and gets her colleagues to adhere to them as well), she can breathe easily again. These conventions cover original and borrowed work, including illustrations, and the appropriate means of obtaining permission when it is needed.

Using Only Original Work

If you always create totally original work, never drawing upon other sources, you will, of course, not infringe upon anyone else's rights. *Original,* as used in this context, means that the work has sprung straight from the creator's own mind and hand to that particular expression and has not been expressed in that particular way (words, visualization) by anyone else—even by that creator. This is the precept that is violated by self-plagiarism, a much misunderstood act. For a work (and particularly a scholarly or scientific work) to be original under copyright law, it must not have been expressed in that way before by anyone. You are implicitly claiming that your work is original and expressed in that way for the first time if you do not include an appropriate citation to work previously expressing the idea or information in the same way (even your own work, in published form or not).

Using Uncopyrighted Work

Another source authors may use without fear of infringement is uncopyrighted work. Uncopyrighted material includes anything that is either fact or in the *public domain.*

- Facts are not copyrightable, so information you obtain from an almanac or from a table of data is not protected by copyright. Professional ethics demand, however, that credit be given for data taken.
- Anything that was once copyrighted but whose copyright has expired or has been otherwise relinquished is in the public domain and not protected by copyright law.
- Many other things are not copyrightable: ideas, concepts, discoveries, processes, procedures, principles, and systems (USC sec. 102b). Works of the United States government likewise are not copyrightable. Those include anything created by United States government employees in the course of their employment. An example might be a handbook written by one or more United States Department of Agriculture employees detailing government-recommended procedures and techniques for avoiding nutrient depletion and soil erosion in crop farming.

It should be remembered, however, that uncopyrighted things may still be protected by patent law (which protects inventions from being manufactured, used, or sold without permission), be considered trade secrets, or be governed by professional ethics.

Knowing the Status of Borrowed Works

If you are drawing upon someone else's work, you must determine the copyright status of that work and adhere to the guidelines established for borrowing from works in that category. This means that you must first determine whether the work is copyrighted.

- **Works published before March 1, 1989,** were required to bear notice if copyright was claimed, though unpublished works were not. But even for published works, the absence of notice does not necessarily mean that the work is in the public domain because the 1976 Copyright Act (effective in 1978) provided remedies that could protect a work even though its notice had been omitted.

- **Works published on or after March 1, 1989,** do not need a copyright notice to be protected according to provisions of the Berne Convention Implementation Act (which resulted in revisions to the 1976 United States copyright law). Incentives remain in the revised law for using the copyright notice, however, and the Universal Copyright Convention still requires it, so publishers will probably continue to use it for most work.

When you are unsure about the copyright status of a work, check on it. Start by writing the author or the publisher, or both. If the work does not give you sufficient information to do this, check the records of the Copyright Office in Washington, D.C. These are open to public inspection, or the Copyright Office will search them for you (at $10 per hour). You could also check the official *Catalogue of Copyright Entries,* which is available in many large libraries (CBE Style Manual Committee, 1983). Because transfers of ownership are often registered, they may be listed in the *Catalogue* also. (The copyright notice does not indicate transfers of ownership.) But, because registration has not been required as a condition of copyright protection since 1978, the fact that you do not find the work listed in these records does not necessarily mean that it is in the public domain. Because the term of copyright protection for pre-1978 works lasted a set number of years, the Copyright Office records can help you determine whether copyrights for these works have expired. From the records, you can determine the original date of copyright and whether or not registration was renewed for a pre-1978 work (the notice does not indicate whether the copyright was renewed).

Copyright has expired in all works published in the United States before September 19, 1906. If the 28-year renewal term allowed by the 1909 Copyright Act was not obtained for a pre-1978 work in the last year of its initial term of 28 years, copyright in that work expired at the end of the first term (also 28 years). In the 1976 Act, Congress added 19 years to the renewal term for works in their initial term on January 1, 1978, giving those works a possible term of 75 years from the date of first publication, if renewal was obtained. Works in their renewal term or for which a renewal application had been filed on or after December 31, 1976, also had 19 years added to their terms. Therefore, you can calculate the latest possible year of United States copyright protection for works published between September 6, 1906, and January 1, 1978, by adding 75 years to the copyright registration date, which should appear on the notice for these works (Johnston, 1979). For example, any works with a United States copyright of 1914 or earlier entered the public domain by the end of 1989, if not earlier.

No such simple formula can be generally applied to works published in 1978 or after because the term for most of these is based on the lifetime of the author. The 1976 Act protects copyrighted works for a single term, the life of the author plus 50 years (for joint authors, this is based on the life of the last surviving author). Exceptions are work made for hire and anonymous or pseudonymous works, which are protected for 75 years from the date of publication or 100 years from the date of creation, whichever is shorter. If an anonymous or pseudonymous author reveals his or her identity to the Copyright Office before this term expires, the term reverts to life-plus-50 years.

If a work was created before 1978 but not published or registered before then, it has at least 25 years of protection under the 1976 Act, and so will be protected at least until December 31, 2002. If it is published by that date, it will be protected through at least December 31, 2027 (Strong, 1982). If the author is still living, the term may be longer.

Adhering to Fair Use Guidelines

If a work is protected by copyright, you may still borrow without permission relatively small portions of it for certain purposes under the *fair use* exemption in the copyright law—provided that you give proper credit to the author. The fair use guidelines are specified in the statute and require consideration of

- The nature of the use to which the borrowed material will be put.
- The nature of the copyrighted work.
- The amount and substantiality of the portion that is borrowed.
- The effect of the borrowing on the potential market for or value of the protected work.

The statute is silent on the relative significance of each of these factors, nor does it prevent courts from considering other factors that may have a bearing on fairness of use. Thus, there is no automatic exception for educational or nonprofit uses. Some tests of these factors have become fairly universally accepted in the courts, however. For example, many courts apply the so-called functional test to determine whether the disputed use is fair according to the first factor listed above: if the underlying and derived work serve the same function, the derived work is probably not a fair use of the first work. In commercial terms, this means that you should never reprint someone else's work so extensively that the "market" for the original product is diminished. Even if a work is not copyrighted, the fair use guidelines are good rules of thumb for the ethical use of borrowed material. So is Johnston's (1979) advice to judge whether your use is "fair" by looking at it "from the other person's perspective" and then considering "whether any likely complaints would make sense to you" (p. 86). For example, if you let a friend pirate a copy of a computer program you own, consider the compensation you would lose from this type of theft as the writer of a program that may have taken years to create.

Avoiding "Derivative Works" Infringement

If your use goes beyond the level of borrowing covered by the fair use guidelines to the point of modifying an earlier work (as is often the case in scientific and technical illustrations or maps), be careful not to create what copyright law calls a *derivative work* without getting permission from the copyright holder. The reason for this is that the copyright holder owns the exclusive rights to derive new creations from the copyrighted one, and if you do that, you are infringing upon his or her copyright. Derivative works are the subject of much controversy in the law, but some guidelines have been at least tentatively established and can guide you in your work.

The basic test for a derivative work is that the new work must not be "substantially similar" to the original work. If it is, it is considered derivative, and the copyright on the original work has been violated unless permission was secured from the copyright owner. Commercial considerations also are usually taken into account by courts in deciding whether a derivative work has been created. On the other hand, even a new work found by the court (or agreed by the parties) to be a derivative work (that is, substantially similar to the underlying work) may not be an infringement if it constitutes a fair use of the underlying work. For example, a derivative work created without permission of the copyright holder of the original work might not be a violation of copyright law if the derivative work is put to an entirely different use than the original work was (the first of the fair use guidelines listed above).

Using Illustrations

One category of technical or scientific work that seems to require special consideration regarding the copyright guidelines is that of illustrations. Rather than being a mere decorative addition to the text, as may be the case in creative works, scientific and technical illustrations may summarize important data, convey models or concepts, or be an integral part of specific instructions. In that sense, they are works in and of themselves. The fair use guidelines of the Copyright Act, therefore, must be considered from this perspective. Borrowing one half-page illustration from a 20-page scientific article may seem insubstantial (applying the fair use guidelines) until you realize that the single illustration you have selected summarizes the entire point of the research. Without specifically mentioning scientific illustrations, the fair use guidelines do take into account the qualitative as well as the quantitative contribution of a given section to the whole work. Because of this, and the implicit ethical considerations, the safest approach probably is to treat each copyrighted scientific or technical illustration as a separate work rather than as only a small part of a larger whole. Hester, Monday, and Snead (1989) have proposed a systematic method for creating borrowed illustrations along with guidelines on when permission to borrow should be obtained.

Obtaining Permission to Exceed Fair Use Limits or Create a Derivative Work

If you want to exceed fair use limits or create a derivative work, then you must obtain permission from the holder of the copyright in the original work by completing the following steps:

1. Determine who owns the copyright.
2. Request permission, in writing, to use the work, and specify what you want to copy or adapt, the use to which you plan to put it, and the sort of credit you plan to give. Because it is sometimes difficult to locate a copyright holder, most publishing houses have established guidelines for what the law calls "making a good faith effort" to obtain permission. Usually this means writing (and keeping copies of) three letters to the copyright holder, spaced at appropriate intervals. If no response is received, you may be considered to have made a reasonable attempt to obtain permission, and can at least make that argument later in court, if a lawsuit results.
3. When permission is granted, adhere to any restrictions specified by the copyright holder and give proper credit to both author and copyright holder. In the noncommercial world (scholarly and educational channels) money may not be an issue in granting permission, though proper credit given to the source is almost always a condition. In the commercial world, the use of the copyrighted material is generally granted only for a fee.
4. Always provide complete, accurate credit for both copyrighted and uncopyrighted material. Use standard reference and citation formats, which can be found in any good style manual, for textual citations, the reference list or bibliography, and for illustration credits. If you do obtain permission to use copyrighted material, you must follow the copyright holder's specifications for providing credit, even if that does not conform to the style you are using in the rest of the document.

HOW TO PROTECT YOUR OWN WORK

Now that she knows some of the complexities of professional standards and copyright law as it relates to technical communication, Jane feels ready to apply the guidelines when she returns to the office. She does think, however, that her information would make a bigger impression on some of the scientists if she can present it in terms of protecting their own work, rather than criticizing them for not respecting the protections afforded other people's. She makes a list of ways in which technical communicators can protect their original work.

Once an original work is *fixed* "in tangible form," it is instantaneously copyrighted under the new law—copyright "is an incident of the process of authorship" (Copyright Office, 1987, p. 3). The copyright immediately becomes the property of the author, and he or she need take no further immediate action to ensure legal copyright protection.

But rarely are things so simple. Consider, for example, that the person who created the original work may not be the author, as in the case of a work made for hire. The employer or client may be considered the author in that situation, and, therefore, becomes the initial holder of the copyright.

If the work remains in the desk or computer or studio of its creator, it is probably safe from infringement. The difficulties arise when the work is published, distributed, or displayed. Although none of these acts will destroy the legal copyright protection, when an author or copyright owner wishes to release the work in some way, it is advisable to take certain precautions: using the copyright notice, registering the copyright, and maintaining high standards for the ethical treatment of other authors' works. The next sections explain how to take these precautions.

Definition of "Publishing"

The statutory concept of *publication* is not limited to formal publishing by a recognized publishing house. Under the copyright statute, publication refers to the release of the work to the public, in any quantity, no matter how small, in any form, no matter how casual. In fact, the work does not have to actually be distributed; the act of offering it for distribution to the public is sufficient to constitute publication (Strong, 1982). For example, if you make handouts available to the public by distributing them at a conference presentation, the handouts are considered published material under the copyright statute.

Circulating copies for limited purposes within a limited group who understand that they are not free to distribute the work to others, however, does not constitute publication (Strong, 1982). Therefore, providing review copies to peers or coworkers would not constitute publication. As a further precaution, some authors include a statement on the title page of review copies, such as, "This copy is for review only. It may not be cited, quoted, or distributed."

How to Use the Copyright Notice

Although the copyright notice is no longer required to protect published work, and was not required by the 1976 Act to protect unpublished work, using it still has definite legal and procedural advantages.

- It is probably the least expensive and simplest way to discourage infringement of your work (Latman, et al., 1989). It conveys useful information to potential users and warns them that the work is "owned."

- The law doubles the statutory damages allowed to holders of registered copyrights and makes recovery of attorney's fees more readily available to them. If copies bear a copyright notice, then, as under the 1976 law, alleged infringers may not claim that the infringement was "innocent" if they had access to the copies bearing notice.
- The copyright notice also provides protection for works used in other countries, a growing concern as we move toward a global economy. Use of the notice is still required for protection by the Universal Copyright Convention. Approximately 20 countries adhere to the Universal Copyright Convention, but not to the Berne Convention, so using notice is still a requirement of protection in those countries (Nimmer & Nimmer, 1989). If you are interested in protecting your work in particular countries, however, you should check on the extent of protection offered and the requirements for obtaining it. Some countries simply do not offer copyright protection for foreign works, and some offer very little (Copyright Office, 1987). The Copyright Office's Circular 38a provides a list of countries that maintain copyright agreements with the United States.

The copyright notice itself consists of three elements: the copyright symbol (©) or the word "copyright" or the abbreviation "copr."; the year of first publication; and the name of the copyright owner. For example, a notice might read:

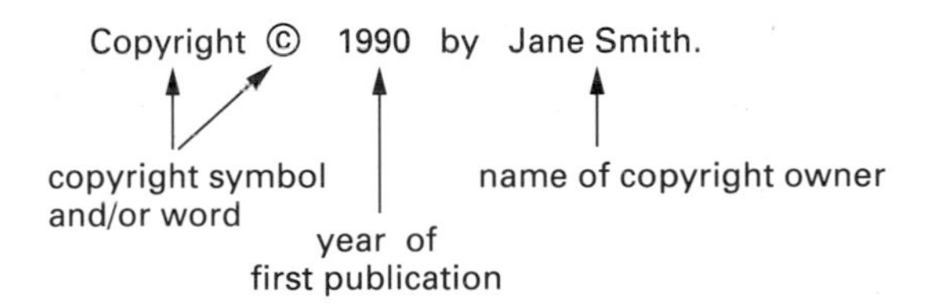

Note the following:

- Many publishers use both the symbol and the word "Copyright" in the notice because the Universal Copyright Convention requires that the symbol be used rather than the word; the word is often included to make the notice clearer.
- The year of publication is important because it is the basis for determining the duration of copyright for works made for hire and for anonymous and pseudonymous works.
- It is a good idea to include the copyright notice on unpublished works to protect them in case of inadvertent publication, even though the notice is not required. This type of notice might read: Unpublished work, copyright © 1990 by Jane Smith.

- If the work consists preponderantly of works of the United States government (which are not copyrightable), the notice must also include a statement identifying either the parts protected by copyright or the parts composed of United States government material.

The law specifies that the notice be placed where it will "give reasonable notice of the claim of copyright." Examples of acceptable positions for the copyright notice in a book or report are

- The title page.
- The page after the title page.
- Either side of the front or back cover.

For works reproduced as machine-readable copies—such as computer programs—the copyright notice requirements differ. For these works, the notice may

- Be affixed on visually perceptible printouts (such as a listing of source codes) with or near the title or at the end of the work.
- Be displayed on the user's terminal when the program is started.
- Be displayed on the user's terminal throughout use of the program.
- Be securely affixed to the program diskette or packaging (Chickering & Hartman, 1987).

Publishers of computer programs often place the notice on the source code, sign on, and packaging (Chickering & Hartman, 1987).

How to Register Your Work with the Copyright Office

Registering your work with the Copyright Office offers advantages, even though you do not have to register it to be protected. Also, once you affix the notice to your work, you are subject to *mandatory deposit requirements*, and a single deposit can satisfy both the notice and registration requirements. Registration

- Establishes a public record of the copyright claim.
- Provides *prima facie* evidence in court of copyright ownership (if the work is registered within five years of publication).
- Is usually necessary before an infringement suit may be filed in court (for United States works).
- Makes it possible for the copyright owner to collect attorney fees and statutory damages as well as actual damages and profits in the case of a court action (if the work is registered within three months of publication).

A work may be registered anytime during the life of the copyright, but certain advantages are lost if the work is not registered within specific periods. Of course, the best time to register a work also depends upon your situation. You may have reason to register the work to protect it even before it is published. If you do, you will need to register it again after publication if you have made any changes in the published version that you wish to protect, or if you want the year of registration to reflect the year of publication. For most authors, the best time to register the work will probably be within the first three months of publication.

An author is not the only person who may submit an application for copyright registration of a work. Others are another *copyright claimant* (defined as a person or organization that has obtained ownership of all of the copyrights that initially belonged to the author), an owner of exclusive rights, or a duly authorized agent of the owner. An applicant does not need an attorney to prepare or file the application; the Copyright Office supplies the forms along with instructions and will also answer questions for applicants. The office also publishes clear, informative circulars on various topics related to copyrights; the Copyright Office's Circular 2, "Publications on Copyright," lists the materials available. Requests for publications or questions should be addressed to the

Copyright Office
Information and Publications Section
LM 455
Library of Congress
Washington, DC 20559

To register a work, send

- The appropriate application form for the work.
- The appropriate deposit requirement.
- The $10 filing fee.

Application Form. The application form must be one from the Copyright Office or an exact clear copy of the form. The form must be completed in black ink or typed. The certificate of registration is reproduced directly from the form, so if the form does not meet these requirements, the office will return it.

There are different forms for different types of work. Technical communicators will probably need one of the following:

- Form TX, for published and unpublished nondramatic literary works (including computer programs and types of technical writing)
- Form SE, for serials (works issued in successive parts numerically or chronologically designated and intended to be continued indefinitely, such as newsletters or journals)
- Form PA, for published and unpublished works of the performing arts (including audiovisual works)

- Form VA, for published and unpublished works of the visual arts (including maps, architectural blueprints, and advertising art, and technical diagrams)
- Form GR/CP, an adjunct application used to register a group of contributions to periodicals (used in combination with Form TX, PA, or VA)
- Form CA, to correct or amplify information submitted to the office in a previous registration

Help in determining which form to use and what to send for deposit requirements is readily available. Chickering and Hartman's book, *How to Register a Copyright and Protect Your Creative Work* (1987), provides copies of the different types of forms and detailed instructions on how to fill them out, along with more general advice and information. The Copyright Office also provides information on filing and selecting the appropriate form (send requests to the Copyright Office, Information and Publications Section to the address provided earlier in this chapter). You may also call 24 hours a day to leave a recorded message requesting forms at (202) 287-9100, or call the Copyright Office Information and Publications Section at (202) 479-0700 between 8:30 and 5:00 (Eastern Time) to ask questions related to filing an application.

Deposit Requirement. You must submit a certain number of copies of the work for the Copyright Office's files; this is called the *mandatory deposit requirement.* One complete copy of the work must be sent for unpublished works being registered; two complete copies of the best edition of works first published in the United States on or after January 1, 1978; two complete copies of the work as first published for works published before January 1, 1978; one complete copy of the work as first published for works published at any time outside the United States; and one complete copy of the best edition of the collection for a work that is a contribution to a collective work and was published after January 1, 1978. The Copyright Office supplies information on mandatory deposit requirements in Circular 7d.

Special deposit requirements apply to the following works:

- Identifying materials such as photographs or drawings are usually required for three-dimensional works.
- Certain categories of work may be exempted entirely from the mandatory deposit requirement, or the requirement may be reduced.
- Actual copies of machine-readable material (such as computer programs) are not deposited; instead, as with three-dimensional works, the Copyright Office requires deposit of identifying material. Several alternatives are allowed to meet the requirement for identifying material for computer programs. For example, you may deposit one copy of the first and last 25 pages of a printed listing of the

source code of the program; depositing a copy on diskette, however, will not meet the requirement because it is machine-readable only (Chickering & Hartman, 1987). If the source code contains trade secrets that you wish to protect, other alternatives are available, including deposit of the first and last 25 pages of source code with the trade secrets blocked out and a written request for special relief. In this type of situation, it may be advisable to consult a copyright attorney about the best alternative (Chickering & Hartman, 1987).

Filing Fee. The $10 filing fee may not be sent in cash. Send a check, money order, or bank draft payable to the Register of Copyrights, and attach it securely to the application.

Application Package. It is important to send the application, fee, and deposit materials in the same package, addressed to the Register of Copyrights, Copyright Office, Library of Congress, Washington, DC 20559. Applications and fees not accompanied by deposit materials will be returned. Unpublished deposit material sent alone will probably be returned also, but published materials sent without the application form and fee will be immediately transferred to the collections of the Library of Congress. Once this happens, the copy may not be used to satisfy the mandatory deposit requirements for registration.

Notification of Receipt. The Copyright Office does not send any notification that the application has been received. If you want such notification, send the materials by registered mail and request a return receipt. The Copyright Office suggests that you allow at least three weeks for the return of the receipt. You may want to have this record of the date that the materials are received because registration becomes effective on the date that the application form, fee, and deposit materials are received by the office in acceptable form, regardless of how long it takes to process the application.

THE BEST PROTECTIONS

The most obvious way in which technical communicators can protect their own work from infringement is to foster, by word and behavior, a high level of professionalism within the field. Always use full, proper citations when using the work of other people, whether it is copyrighted or not. Be generous in allowing colleagues credit for work they have done, but don't abuse coauthorship by listing "honorary" authors. Be a stickler for detail and precision in reference formats, footnote style, illustration captions, and the like. Err on the side of caution in deciding when a "borrowing" fits the fair use guidelines. Do not succumb to inertia—if your office does not have a strict policy of obtaining permissions for borrowed works such as copyrighted technical diagrams and maps, take the initiative to get out a style manual and follow the instructions to write a letter requesting permission. Do not accept the rationalization that the owner probably will not sue you or perhaps even discover the borrowing. The copyright laws are based on the belief that authors of works are entitled to benefit from their creations. Our professional ethics support the conviction that, beyond legalities and penalties, *creators* ought to be recognized for their contribution and that no one has the right to claim, even by silent implication, another's work.

SUMMARY

This chapter has examined the rights and obligations inherent in the use of technical, scientific, and other material commonly used by technical communicators. Although the field of copyright law and professional ethics may seem overwhelmingly complex, technical communicators need to become at least somewhat familiar with the basic principles of the legal and ethical use of original materials, whether they be books, illustrations, or computer programs.

- Two interdependent sets of rights, obligations, and guidelines govern the use of original works of authorship generated by people in the United States: the laws of copyright, which protect ownership, and the rules of professional ethics, which protect originality.
- Technical communicators can avoid infringing on other authors' legal and ethical rights by
    - Knowing the status of borrowed work and taking appropriate actions, including obtaining permission to use borrowed material that is copyrighted or otherwise proprietary
    - Adhering to the copyright law's fair use guidelines
    - Obtaining permission to create derivative works
    - Always giving credit to the original author when using borrowed material
    - Treating illustrations with special care
- Technical communicators can protect their own original works by
    - Affixing an accurate copyright notice, in the proper position, as soon as the work is created. The copyright notice includes the copyright symbol (or the word "copyright"), the year of first publication, and the name of the copyright owner.
    - Registering the work with the Copyright Office within three months of publication. The registration application includes the correct form, the mandatory deposit, and the $10 application fee.
- Finally, technical communicators can maintain a high level of ethical standards in the field by respecting other authors' works and by adhering to ethical standards in publishing their own works.

LIST OF TERMS

author In copyright law, the first owner of the copyright in any kind of work (does not refer simply to writing). This would be the creator *unless* the creation was a work made for hire; in those cases, the employer or commissioner is the "author." Also see *creator.*

coauthor Person listed as an author of a work who directly and substantially contributed to the work and is able and willing to take public responsibility for its content and quality.

common law Law based on precedents established in courts rather than on legislative action.

copyright claimant Person or organization that has obtained ownership of all of the copyrights that initially belonged to the author, an owner of exclusive rights, or a duly authorized agent of the owner.

copyright notice Statement providing the copyright holder certain benefits if it is positioned on a work so that it will "give reasonable notice" that copyright is claimed, and includes (1) the copyright symbol, the word "copyright," or the abbreviation "copr."; (2) the year of first publication of the work; and (3) the name of the copyright holder.

creator Person who created the work as a product of his or her own mind and hands (sculptor, composer, writer).

derivative work Work substantially based upon one or more preexisting works (for example, a translation, abridgement, screenplay, musical arrangement, art reproduction).

fair use Limited use of a copyrighted work for certain purposes without the copyright owner's permission.

fixed Put in tangible form so that others can perceive it.

honorary author Person listed as a coauthor who made no essential intellectual or technical contribution to the work. The practice of naming honorary coauthors is considered unethical.

mandatory deposit requirement Certain number of copies of a work sent to the Copyright Office as part of the requirements for registering a work; kept on file.

originality In copyright law, the quality in a work of authorship of being the product of the creator's own mind and hands. In scientific publishing, the quality in a work of authorship of the concepts, theories, findings never before having been released to the public (published for the scientific community).

ownership In copyright law, the rights given by a person or persons owning one or more components of the copyright in a work of authorship (may or may not be the creator or author; may be someone else). The owner, or entity, may not own all the rights and may not be the only owner of a particular work. This entity is also called a nonexclusive owner or holder of nonexclusive rights.

plagiarism Act of presenting someone else's words, ideas, or data as though they were your own.

prima facie (evidence) Evidence that would, if uncontested, establish a fact or raise a presumption of a fact.

public domain (works) Works that are not copyrighted and may be used without restriction.

publication Act of offering a work to the public (not necessarily for sale; not necessarily distribution).

self-plagiarism Act of publishing information you have already published without properly citing the previously published work and obtaining permission from the copyright holder to reprint the material.

work made for hire Work prepared by an employee within the scope of his or her employment or a work specially ordered or commissioned if the parties have specifically agreed in writing that the work shall be considered a work made for hire and other specific requirements are met.

REFERENCES

Abelson, P. H. (1982). Editorial: Excessive zeal to publish. *Science 218,* 6251.

CBE Style Manual Committee (1983). *CBE style manual.* (5th ed.). Bethesda, MD: Council of Biology Editors.

Chickering, R. B., & Hartman, S. (1987). *How to register a copyright and protect your creative work.* New York: Charles Scribner's Sons.

Copyright Office (1987). *Copyright basics,* circular 1. 181-532/40,027. Washington, DC: U.S. Government Printing Office.

Hester, M. C., Monday, J. L., & Snead, J. I. (1989). Documenting illustrations. *Technical Communication 32,* 102–113.

Huth, E. J. (1982). Authorship from the reader's side. *Annals of internal medicine 97,* 613–14.

Johnston, D. F. (1979). *Copyright handbook.* New York: R. R. Bowker.

Koshland, D. E., Jr. (1987). Fraud in science. *Science 235,* 141.

Latman, A., Gorman, R. A., & Ginsberg, J. C. (1989). *Copyright for the nineties: Cases and materials.* (3rd ed.), Charlottesville, VA: The Michie Co., Law Publishers.

Nimmer, M. B., & Nimmer, D. (1989). *Nimmer on copyright: A treatise on the law of literary, musical, and artistic property, and the protection of ideas.* New York: Matthew Bender.

Smith, A. J. (1985). Scientific fraud probed at AAAS meeting. *Science 228,* 1292–1293.

STC (1982). Code for Communicators. Washington, DC: Society for Technical Communication.

Stewart, W. W., & Feder, N. (1987). The integrity of the scientific literature. *Nature 325,* 207–214.

Strong, W. S. (1982). *The copyright book: A practical guide.* Cambridge, MA: MIT Press.

Zen, E-an (1988). Abuse of coauthorship: Its implications for young scientists, and the role of journal editors. *Geology* (April), 292.

INDEX

D